DERBYSHIRE CAVEMEN

*Best wishes,
Steve Cliffe*

DERBYSHIRE CAVEMEN

STEPHEN CLIFFE

AMBERLEY

*This book is dedicated to those
ancient people, our remote ancestors,
whose land we hold in trust
and whose spirit, whether we know it or not,
still influences us.*

*Also my Mother, Father and Brother,
without them it would never have been written!*

First published 2010

Amberley Publishing Plc
Cirencester Road, Chalford,
Stroud, Gloucestershire, GL6 8PE
www.amberley-books.com

Copyright © Stephen Cliffe 2010

The right of Stephen Cliffe to be identified as the Author
of this work has been asserted in accordance with the
Copyrights, Designs and Patents Act 1988.

All rights reserved. No part of this book may be reprinted
or reproduced or utilised in any form or by any electronic,
mechanical or other means, now known or hereafter invented,
including photocopying and recording, or in any information
storage or retrieval system, without the permission in writing
from the Publishers.

British Library Cataloguing in Publication Data.
A catalogue record for this book is available from the British Library.

ISBN 978 1 84868 156 9

Typesetting and Origination by Fonthill.
Printed in the UK.

CONTENTS

	INTRODUCTION	7
1.	SETTING THE SCENE	10
2.	NEW WORLDS	19
3.	WONDERS OF THE PEAK	44
4.	ANTIQUARIES AND BONE CAVES	58
5.	GROTTO OF THE GODDESS	69
6.	CASTLES AND CAVERNS	88
7.	ART AND MAGIC?	111
8.	GAZETTEER OF CAVES	129
	CONCLUSION	180
	MAPS	182
	BIBLIOGRAPHY	187
	ACKNOWLEDGEMENTS	189
	INDEX	190

Ilam Rock, Dovedale, has a cave at its base, and is an isolated column of limestone.

INTRODUCTION

Any region of Britain with suitable geology for caves might be selected to give a localised example in detail of the habits of our cave-dwelling ancestors. So, naturally, I chose to examine the area best known to me. Antiquaries have been rooting about in the Peak District of Derbyshire, looking for evidence of former inhabitants, for the past three centuries at least. Few places in Britain have such an abundance of opportunity to delve into the very ancient past.

Because of its interesting geology, the Derbyshire Peak's karst limestone, which in China has eroded into great pinnacles of rock, here produced deep gorges and high, exposed cliffs, with some areas like Dovedale showing miniature examples of isolated columns of rock. Gouged by meltwaters of successive ice sheets, the semi-porous limestone formed caves, which early man utilised, initially for shelter in his hunting and gathering operations, and later as sepulchres for the dead.

The southern central plateau of the Peak, known as the 'White Peak', is composed of this material, the legacy of millions-of-years-old, dead sea-creatures and algae, shot through with volcanic vents and mineral veins which have been exploited by miners since Bronze Age times. Straddling the borders of Derbyshire and Staffordshire, the limestone has yielded vast quantities of lead – and once boasted at Ecton, in the Manifold Valley, the largest copper mine in the world.

To the north, the 'Dark Peak', a grit and sandstone wilderness intersected by river valleys, contains the highest hills, now peat-covered moorlands, but in Mesolithic (Middle Stone Age) times an extensive birch forest filled with game, where, at the level of the sand beneath the peat, hand-axes, flint arrowheads, and spear tips are continually discovered as evidence of the millennia of human activity in a now vanished, once verdant, landscape.

These ancient tools are also found and collected by farmers in both the White and Dark Peak, and quarry owners stripping the topsoil for quarrying have uncovered still more – many of which can be seen in local museums, while others are in private collections. The oldest flint implement, a hand-axe, dates back roughly 200,000 years, and is displayed at Buxton Museum.

There are an estimated 260 surface-accessible, natural caves in the White Peak, all of which will have been explored by humankind at one time or another and more caves, which have been lost or remain hidden with blocked entrances. Excluding later mining activity, which has produced many miles of tunnels, nearly 100 of these caves and caverns have yielded human remains, and evidence of early human activity stretching back to Neanderthal times and beyond: up to 200,000 years ago, before 'modern' humans arrived at Creswell Crags, and long before the final cold period of the ice age!

By contrast, the gritstone of the Dark Peak does not provide extensive cave shelters, but a few exist – notably at Ludchurch Cavern in the Dane Valley, Robin Hood's Cave on Stanage Edge, and a cavern or rift at Kinderlow End, Kinder Scout, the Peak's highest and bleakest hill. An interesting gritstone shelter, the Hermit's Cave at Cratcliffe Tor has a remarkable medieval crucifixion scene carved into the rock, and there are Druidic associations with caves at Rowter Rocks, across the valley. Much older prehistoric rock art exists in a magnesian limestone gorge at Creswell, on the borders of Derbyshire and Nottinghamshire, the first place such art was discovered in Britain, in caves with an abundance of ice age remains.

Most likely, early man built shelters of wood and thatch, or hides stretched between poles or animal bones, in areas where caves were absent, though early concentric stone structures have been uncovered at Deepcar with the remains of a flint knapping industry on the Derbyshire/Yorkshire borders. Also, foundations of an early village of what are believed to be Mesolithic roundhouses were discovered in fields just outside Buxton.

This book is partly an introductory guide to the sites, including the large show-caves at Buxton, Castleton, and Creswell, but also examining lesser-known systems like the ancient red sandstone diggings at Alderley Edge on the fringe of the Peak in Cheshire, where Roman and Bronze Age remains have been found. Scores of other caves have been excavated, yielding human bones and archaeology, but not always as startling as the discovery, by the Pegasus Caving Club, at a newly opened cave on Carsington Pasture, of a stalactite growing through a human skull…!

Visitors to caves are reminded that these sites are an important part of Natural England's places of special scientific interest, and may be on private property. Many are designated ancient monuments, and unauthorised interference is both illegal and can ruin and deface what has been preserved for millennia, which we the living have a duty to protect

Rocky shelf on Stanage Edge, near Hathersage, conceals entrances to Robin Hood's Cave.

for many others to enjoy. Those wishing to explore caves should belong to a recognised caving club, and do so at their own risk with adequate insurance against accident. Finally, this is not a scholarly treatise, but an introduction for the intelligent casual reader to the fascinating but little-known world of cave folklore and archaeology.

Stephen Cliffe, 2010

1
SETTING THE SCENE

GEOLOGY

Limestones of the White Peak were laid down in tropical seas, 300 to 330 million years ago, when what is now Britain lay on the equator. Marine algae, crinoids, and a variety of shellfish, but comparatively few corals, contributed to create Derbyshire's limestone in shallow lagoons surrounded by tropical atolls, like the South Sea Islands of today.

The earliest limestone sits on slaty rocks of Ordovician age, which now lie at a depth of almost a mile – based on a borehole drilled at Eyam. Movement of the Earth's crust lifted the limestone above sea level, causing erosion of the upper beds. Small volcanoes erupted, spreading basalt and lava flows, known to Derbyshire miners as toadstone. Explosive eruptions yielded showers of volcanic ash, or tuff. A number of extinct volcanic vents are still exposed in the limestone.

Massive river deltas were created by changes in geography, and deposited sedimentary particles of eroded mountains, from the Scottish Highlands and Greenland, as sand and mud all over the limestone. This, in turn, was covered by swamp-like forests which became the coal measures. Further movements in the Earth's crust caused an up-arching of the region by east/west compression. The resulting dome was eroded, exposing the ends of the bedding planes as ridges of alternate types of rock – sandstone and grits to the north, as the Dark Peak, and on the eastern and western edges the coal measures as the former topmost layer on either side in Lancashire and Yorkshire, and between beds of sandstone and shale, while the central southern limestone at the deepest layer was exposed as the White Peak. Fractures in the limestone had become filled with mineral-rich liquids under pressure, depositing as veins of lead, copper, quartz, and fluorspar.

ICE AGES

It is difficult for us to imagine that for the last 700,000 years, Britain has spent more time joined to the continent as a peninsula of Europe than it has as an island. In that time, there may have been six successive stages of detachment – sometimes with sea levels much higher than today, and little evidence of a human presence in the island, and as many as seven glacial maximums, when the cold was intense, and ice sheets advanced, then retreated, in cycles lasting between 40,000-100,000 years each.

The ice-age epoch of geological time is referred to as the Pleistocene, and is estimated to have occupied, altogether, a period of 1.8 million years. Ice sheets covered much of Europe and North America in the coldest glacial stages, broken by warmer interglacial respites, when the glaciers retreated northwards. Two major ice advances across Britain were in the Anglian, which reached the Thames and the Bristol Channel upwards of half a million years ago and, four glacial maximums later, the Devensian, which slid its glaciers around Derbyshire Peak, but may not have actually covered it completely, about 50,000 years ago. Early humans had already appeared in the area, leaving a scatter of tools and butchered animal bones by then.

Path alongside a drystone wall leads to Windy Knoll Cave, below Mam Tor, Castleton.

We are presently living in one of the longer warm intervals, or interglacials, which began about 13,000 years ago, with the start of the Holocene period. It is against this fluctuating climatic background that the human occupation of the Derbyshire caves has to be understood. A variety of theories seek to explain the onset of ice sheet advance. They vary, from interstellar dust particles reducing heat received from the sun, to ocean conveyor current fluctuations cutting off warmth, and the effect of axis wobble and changes in the earth's orbit around the sun.

EARLY HUMANS

The earliest human remains found in Britain are a half-a-million-year-old shinbone, from a quarry at Boxgrove in West Sussex. But more recent discoveries of flint flake tools at Pakefield, in Suffolk, have pushed the first date for occupation of Britain by hominids to 700,000 years ago, when the climate was said to be pleasantly Mediterranean in temperature[1]. A honey-coloured hand-axe of flint is the oldest definitely manmade implement discovered so far in the Derbyshire uplands. Found at Hopton, near Wirksworth, it is estimated as at least 200,000 years old, and is part of an exhibition at Buxton Museum tracing the local development of early man.

Acheulian ovate hand-axe was used by prehistoric Derbsyhire people 200,000 years ago!

---

1 On 7 July 2010, Chris Stringer of AHOB project announced they have discovered 70 flint tools at Happisburgh in Norfolk from 800,000 years ago. This pushes the earliest human evidence in Britain back a further 100,000 years.

A flattened oval shape tapering to a point, with a slightly twisted cutting edge, it is said to show the classic style of the Acheulian, in the mid-Palaeolithic (Old Stone Age) period, and may have been dropped by a hunter during the second of three warm intervals in the various onsets of ice-age. It contains cracks produced by the intense cold of, perhaps, the last two glacial maximums, and may have been washed to its final position by glacial melt-waters as the ice retreated. However, its style is quite similar to the much older tools uncovered near the remains of the earliest human ever found in Britain, at Boxgrove, Sussex, estimated to be up to half a million years old!

It is thought that the hand-axe makers of the Acheulian operated in large groups, in open grassland habitats, with greater opportunities for social intermixing, and more possibilities for learning complex tool-making, which could be passed on from generation to generation. Forested margins were favoured by early hand-axe makers, and at Hopton, and nearby Carsington Pasture, the steep slopes of the limestone grassland sweep down to what may have been a thickly wooded valley on the margins of the White Peak. In addition, this last outcropping of limestone on the edge of the rolling Midland plain offers refuge, in the form of caves. Carsington Pasture Cave has yielded much interesting archaeology in a Time Team dig (more of which later). Hand-axes of similar type have been uncovered in lowland gravel terraces of the Trent and Dove valleys near Derby, all showing signs of having been rolled by glaciers.

However the nearest human *bones* from this early period were found at Pontnewydd Cave, St Asaph, North Wales, where twenty human teeth from five early individuals, and some jawbone, date from over 225,000 years ago, while heated pebbles and hand-axes have been dated to 250,000 years

Fragment of jaw and teeth from Pontnewydd Cave are the northernmost early human remains.

old. But, between the deposit of the Hopton hand-axe, and the coming of the Neanderthals at Creswell Crags 55,000 years ago, no further evidence of human activity has been uncovered. In fact, there is very little sign of people in Britain from about 160,000 to 60,000 years ago, partly because of intensely cold periods, partly because Britain was again subdivided from Europe by very high sea levels. Nomadic early humans and the migratory animals they hunted, driven across the land bridge to more southerly ice-age refuges by the advancing glacial chill, would have been unable to return, after warmer conditions raised the seas again.

Such unlikely animals, from our point of view, as lions, hyenas, and hippopotami had already left their remains in Derbyshire, Staffordshire, and Yorkshire caves during warm phases of the previous interglacial. About 60,000 years ago, animals of the mammoth steppe re-appeared in Britain, as sea levels fell with the onset of another cold period, producing dry continental grassland supporting woolly rhinoceros, wild horse, spotted hyena, bear, and wolf. It is believed that Neanderthals migrated with the animals, bringing their distinctive style of Mousterian flint flake tool industry. One of their hand tools, a grey flint side-scraper was recovered from inaccessible Ravencliffe Cave, high in the cliffs near Cressbrook in the Derbyshire Dales. It was found with an assembly of archaic animal bones, including mammoth, woolly rhinoceros, reindeer, and bear. Quartzite tools and bone splinters of similar age were also found at Harborough Cave near Brassington.

Mousterian side-scraper from Ravencliffe Cave at Buxton Museum.

Ravencliffe Cave, high in cliffs above Cressbrook Dale.

NEANDERTHALS

Controversy still rages as to whether or not these strong, resourceful, survivors of hundreds of thousands of years of changing climatic conditions, flora and fauna, and land geography which sometimes had Britain joined to the continent and sometimes not, were our direct ancestors. What little DNA survives in Neanderthal bones has shown quite different sequences from modern humans. An ongoing genome project hopes to prove that, in part, we may at least be cousins.

One thing is certain. We modern humans have not been around for as long as they were carving a living from the wild beast-filled terrain! Neanderthal as a species of human was first identified as appearing from 200,000 and living up to 30,000 years ago. They may have been descendants of *Homo Erectus* groups who migrated to Asia from Africa 1.8 million years ago. Neanderthal remains have been found throughout Europe, and as far as central Asia, with distinct identifying characteristics. They were very strongly made, with sturdy bones and heavy muscles, used to very gruelling tasks, but with a larger braincase and lower stature than the earliest specimens of modern human, who superseded them about 29,000 years ago.

Neanderthals once inhabited the Derbyshire uplands.

Their heads gave the impression of a jutting face with a large nose and powerful jaws and teeth, heavy brows, a hollow back to the skull, for the attachment of strong neck muscles, and large wide fingers and toes, used for violent gripping in tussles with animal prey.

These fascinating people first appeared in Britain about 60,000 years ago, although even older peoples had been around long before that. The discovery of Mousterian hand-axe and flint flake tools, with mammal bones of the 'mammoth steppe' variety suggest Neanderthals migrated into Britain when sea levels were seventy metres lower than today, following the animals. Mousterian technology was a type of stone tool manufacture of Middle Palaeolithic culture, very much associated with Neanderthal activities. These were the people who appeared, and left their evidence at Ravencliffe Cave in the Peak, and at Creswell Crags on the Derbyshire/Nottinghamshire borders.

At Creswell, Pin Hole Cave, Robin Hood Cave, Church Hole, and Mother Grundy's Parlour all held stone tools left by them. Neanderthals also visited Ash Tree Cave, in a small valley called Burnhill Grips, two miles north of Creswell. They lived in small, mobile groups, using the caves and rock shelters as temporary camps, and often competing with other carnivores – hyenas, lions, and bears – for the right to occupy them! Their prey included reindeer, bison, and mammoth, which may have been ambushed at waterholes, or deliberately driven over cliffs. Healed wounds on Neanderthal bones show many injuries sustained around the head, neck, and shoulders, suggesting close up struggles with wounded herbivores, in which the hunter may have mounted his prey like a rodeo rider, and thrust downwards for the final kill, sometimes being injured by horns or tusks as the dying animal tried to shake off its human attacker. Well-developed muscle attachments on Neanderthal shoulder bones support this theory, particularly for the downward thrusting muscles. Females show similar injuries, and activities between males and females do not appear to have been segregated, the females probably hunting with the males.

The Neanderthals were at Creswell by 55,000 years ago, and made use of the local quartzite stone for on-site manufacture of tools, probably bringing sets of superior flint tools with them from areas where this material was available. Apart from spears and javelins, hand-axes and scrapers would have been needed for butchery and cleaning fat from hides, and woodworking to make handles and bowls of hard wood, like juniper. Anthropologists believe that Neanderthals were highly intelligent and resourceful, the human survival experts of the Ice Age world – not the apelike brutes of popular misconception.

They were similar to us, in that they controlled fire, and made tools. They hunted in organised and very mobile groups, and probably developed vocal skills to direct one another while working in concert. Neanderthal children grew up much quicker than our own. Life expectancy was just thirty-five for adults, whereas early modern humans could expect to reach fifty, bar accident. Modern human children were nurtured for longer (Neanderthal children would have been experienced hunters by the age of twelve). Neanderthals looked after the old and infirm. They buried their dead in a flexed foetal position, or with a hand raised to the face, and deposited simple grave goods like tools, or pieces of animal food. There is evidence of de-fleshing after death (from cut marks on bone), with some cases of bone smashing to extract marrow, suggesting a possible ritual

Creswell Gorge during a glacial stage, by Robert Nicholls. © www.paleocreations.com

Neanderthal group around a campfire in cave at Creswell. A guard wards off hyenas. Painting by artist Robert Nicholls. © www.paleocreations.com

cannibalism, perhaps to absorb the spirit of the deceased. Cannibalism, of course, is not unknown among modern humans.

Mysteriously, Neanderthals disappear from the archaeological record at about the time that modern humans appeared in Europe and the Middle East. It has always been suggested that they were out-competed, and became extinct about 30,000 years ago. Now, some attempt is being made to suggest that a limited absorption into the modern gene pool may have taken place.

2

NEW WORLDS

MODERN HUMANS

The ancestors of modern humans originated in Africa, over 190,000 years ago. They migrated about 70,000 years ago towards Europe, via the Middle East, and were at one time living alongside Neanderthal communities, in one case in caves on opposite sides of a mountain in what is now modern Israel. Remains of modern humans were first identified at a rock shelter in south-west France in 1868, and named Cro Magnon after the place.

Cro Magnon man had a rather different appearance from Neanderthal, and was not unlike a modern European. Their steep foreheads and less pronounced brows, with square eye sockets and a flatter face, tucked in beneath a high rounded braincase with a more prominent chin, but much smaller nose and mouth, gave a comparatively genteel impression. The limb bones were longer and less sturdy, but the height was greater than Neanderthal and even taller than the average today. The brain size, from cranial capacity of skulls, was also greater than the modern average, but curiously less than that of the Neanderthal!

Modern humans entered Europe about 40,000 years ago, and began to replace Neanderthal populations. Their culture exhibits a far greater complexity of technology, supply of raw materials, animal husbandry, social organisation, and even art and music which is as yet unknown among Neanderthal groups. There was a greater use of narrow blades, which could be fitted to projectile weapons, often along a groove, giving an extended cutting edge. A variety of scrapers, borers, piercers, and chisels entered the toolkit. Bone, antler, and ivory points for javelin shafts were manufactured which don't appear among Neanderthal remains. There is evidence of spear throwers, and even a form of boomerang. Animals could be more selectively culled, and even seasonally herded. Food storage pits seem to have been systematically managed. Hearths with complex stone arrangements for food preparation, melting resin, and baking clay first

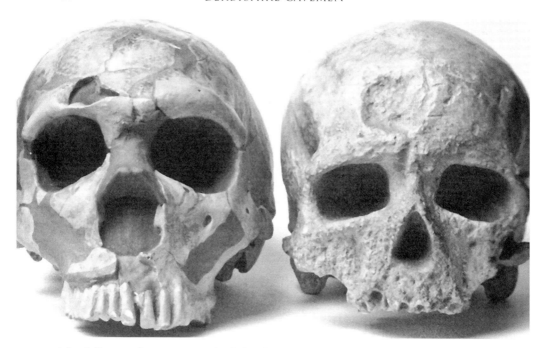

The differences between Neanderthal (left) and Cro Magnon (right) are self evident in these skulls found in France.

Early cave burial, using red ochre powder sprinkled on the corpse, in this painting of Paviland Cave by artist Gino D'Achille.

appear, replacing the simple fire pits of the Neanderthals. Language may have developed in complexity, along with social co-operation between separate groups. Figurines were carved, and moulded from clay, and the first rock art appeared as engravings, or with the use of pigments. Anthropologists have suggested that these became emblems – binding together associated groups. Musical flutes, or whistles, of bone and ivory have been found, and simple lute-like bows have been depicted.

Burials were different from those of Neanderthals, with graves containing single, double, and group inhumations, with the dead lying full-length, dressed in clothing, often woven with shells or ivory beads, which has rotted, leaving only the lines of beads, and implements or tools of stone, placed as grave goods. Red ochre, ground into powder, is found sprinkled on the bones, staining the grave and its contents with a symbolic bloody hue. Caves were used as sepulchres, but some graves have been found in the open. Burials in Britain from this period are incredibly rare.

Pin Hole Cave, at Creswell Crags, yielded a few broken tanged spear points of the Gravettian period of about 29,000 years ago – showing early modern people visited the region before the last glacial maximum set in. Nets for catching smaller prey were in use by this period, and dogs may have already been domesticated as hunting animals.

Early modern humans, hunting with dogs at Creswell and using traps. Painting by artist Robert Nicholls. © www.paleocreations.com

These hunters would have been moving over a bleak tundra landscape, composed of sedge grass and moss, at the northern edge of their territorial range, following reindeer and other cold-resistant herbivores. When the climate became unbearably icebound, about 22,000 years ago, Britain was abandoned for the next 14,000 years!

In this last ice advance, sheets up to one mile thick extended across, and around, the Pennines. Icecaps advanced through the hills as far as Chapel en le Frith, and at Roosditch near Whaley Bridge, Derbyshire, and in a similar feature at West Gate, Lyme Park, Cheshire, glacial run offs were created, and banks of glacial gravel were dumped , creating natural features. Creswell, though ice-free, was within a few miles of the glaciers, and was experiencing freezing polar desert conditions. The cold was so intense that roofs of caves shattered, and thick layers of angular debris built up on their floors. Modern Human hunters had retreated before these inhospitable conditions, occupying refuges on the borders of North Eastern Spain and South Western France, and in Western Ukraine. It was from these ice age refuges that northern Europe and Britain were eventually revisited by some of our direct ancestors, 13,000 years ago.

Venus figure of the ideal prehistoric woman – well-fed to survive the famine times, holding a horn of plenty.

DNA EVIDENCE

Over the last decade, there has been more research into the genetic evidence of ancient remains, thanks to advances in DNA analysis. It is now possible to examine gene sequences from fossilised bone thousands of years old. Where archaeologists and anthropologists once estimated ages from radiocarbon dating and artefacts, a clearer picture of the descent of man is becoming available, via analysis of human relatedness.

It is now possible to say, by observable mutations in DNA sequences, that the common ancestor of Modern Humans and Neanderthals existed about 700,000 years ago. For thousands of years before the last ice age, Neanderthal and Modern Humans lived in Europe alongside one another. Did they interbreed? The answer to that question is as yet unknown. Analysis of Neanderthal bones show that they were 99.5 per cent identical to us genetically, but that critical 0.5 per cent made all the difference in the survival game. Our Modern Human ancestors started their long journey out of Africa 190,000 years ago – the ancestors of Neanderthal somewhat earlier, but by the start of the last phase of the ice, they vanish from the archaeological record. Modern Humans who came to repopulate Britain after the last glacial maximum had survived in an ice-free refuge in what is now the Basque country of north-east Spain. Here, and in south-west France, it is estimated from remains of their implements that 2,000-3,000 people were able to exist in a life-sustaining microclimate there.

Stephen Oppenheimer and other genetic researchers have claimed that modern Britons, living in small to medium sized towns and villages, with long-established local ancestry, can be shown by genetic analysis to be overwhelmingly descended from these returning ice-age hunter gatherers, who came to Britain between 13,000 and 10,500 years ago. This applies to England (and other parts of Western Europe), as well as the areas traditionally assigned to Celtic cultures. In fact, rather than speaking a Celtic language, these first natives of Britain probably spoke a forerunner of the unique Basque tongue. Many Britons, therefore, exhibit a preponderance of a strongly Basque genetic haplogroup R1b – a statistical incidence rising to a high of 96 per cent at Llangefni in Anglesey, while Ireland is considered to show the highest overall concentration of ice-age hunter-gatherer genetics, because of lesser subsequent inward migration.

These assertions were independently supported by the analysis of bones from Gough's Cave, at Cheddar, Somerset, where a 9,000 year old complete skeleton was found, on analysis, to have bequeathed similar genetic sequences to people still living in the village, including the headmaster of the local primary school.

Oppenheimer's investigations seemed to show that contrary to historical views, subsequent invasions by Celts, Romans, Anglo-Saxons, and Vikings have added a mere 5 per cent each to the genetic heritage.

THE FIRST BRITONS

Why do we like to roam about the hills and vales, exploring byways, usually in small groups? Why is fishing so popular in Britain? Why do family outings persist in the tradition of collecting nuts and berries? My own idea is that the bracing open hills of Derbyshire, 'so free', and similar places in our native land satisfy an inner urge to wander, strongly felt in the hearts of the descendants of the ancient Britons. This is the happy hunting ground, not the Elysian fields of Greece, but windswept hillsides with the clouds racing overhead – and what joy to catch sight of a herd of deer! Some of us even slither and slide, and thrust our bodies into wet, rocky crevices in the bowels of the Earth – but that is a step too far for many. This is to enter the realm of the distant ancestors.

I remember as a small child walking along a cliff-fringed beach in Cornwall with my father. Mysterious and intriguing caves loomed darkly in the twisted rock, and perhaps I asked to explore one of these. Holding my hand, he marched boldly into a lofty cavern with a level sandy floor. At some point the passages divided, and I was suddenly afraid to go on.

Cheddar Man, at over 9,000 years the oldest complete British skeleton, still had a descendant in the village!

The hills so free – above Ravencliffe Cave towards Wardlow Hay and Cressbrook Dale.

An unreasoning, instinctive fear of something lurking in the darkness had taken hold. In the days of our remote ancestors, such fears would not have been groundless. Hyenas, bears, and big cats liked to den in caves, and did not take kindly to human visitors! The cave returned to me in my dreams for years to come with an awesome power. Swiss psychologist Carl Jung has described the cave as one of the common archetypal images ingrained in the unconscious mind of all modern humans, with a variety of potent symbolic meanings which may appear in dreams – an imperfectly understood, but normal psychic process. In the depths of caves late in the ice age, early people, with primitive lamps, burning tallow fat, continued an artwork tradition of depicting animals and strange hieroglyphic symbols in shallow pools of light they had created in the darkness. They used burins – simple sharpened scraper tools of stone for engraving, with charcoal for broad black lines, while colour was added from red ochre and vegetable dyes.

Those artists were the natural shamans of the group – the intercessors between the world of spirits, the formulation of ideas, and the physical, material world of solid rock, tools, and implements. 'In trance or hallucination a shaman might pass to the spirit world and adopt the mantle and powers of the revered animal,' says an expert on rock art, Andrew Lawson. 'The tale would be graphically illustrated and belief in its message instilled. Paintings of beings half animal, half human may represent shamans passing from the real to the spirit world.'

Shamans, in trances induced by rhythmic dancing, drew spirits on and from the cave walls.

Rhythmic dancing, and the eating of hallucinogenic herbs or mushrooms, may have aided the trance inducing process, helping artists depict the entopic (animal/ human) forms and hieroglyphs, which are said to flash upon the inner eye in the darkness.

Cave art has been traced from its earliest traditions, appearing about 30,000 years ago, to the later cave art of about 10,000 years ago, at the end of the ice age. The tradition seems to coincide with the arrival of Modern Humans. Until recently, it had been thought that there was no cave art in Britain, but with the discovery of a series of engravings of animals at Creswell Crags in 2003 at Robin Hood Cave, Church Hole, and Mother Grundy's Parlour this idea was overturned. Already at Creswell, examples of portable art had been found, dating to 12,500 years age. These included a human figure inscribed on the end of a woolly rhinoceros rib bone, an inscribed ivory javelin point, and an early horse engraved on a piece of bone, found by Prof. William Boyd Dawkins in 1876.

Most cave murals in France and Spain were depicted around the entrance and just inside, where natural light filtered in, but as the last ice age progressed, deeper, darker passages were utilised, as it was warmer underground. Since the latest discoveries at Creswell, an engraving of a 'mammoth' has come to light at Gough's Cave in Cheddar Gorge, and there are probably others in Britain, which may be hidden by debris in blocked passages, like some of those discovered in European caves. Finding cave art requires an expert eye and special lighting.

Horse's head engraved on a rib bone - found at Creswell by Boyd Dawkins.

These early Britons used a type of flint blade with a blunted back, like a modern penknife. Similarities between specimens found at Creswell Crags, Gough's Cave, Somerset, and Kent's Cavern, Devon, suggest the same groups used all three sites, and give some idea of the mobile range of these early hunters. Some evidence of cannibalism, from 12,000 years ago, exists at Gough's Cave. Domesticated dogs may have helped mobility by use in carrying equipment, as well as tracking game. In the final cold stage, reindeer had started to migrate to Britain again, calving in the spring and heading back towards Germany, across what is now the North Sea, in the summer and autumn. Bones of reindeer driven into a bog at Stellmoor, Germany, were transfixed with pinewood arrow shafts and flint arrowheads, proving that archery was now part of the hunting arsenal of modern humans.

In the Manifold Valley, three caves – Thor's Fissure Cave, Elder Bush Cave, and Ossom's Cave – have each revealed the characteristic thin, flaked-flint tools of the Later Palaeolithic (Old Stone Age) people, and, curiously, a bison bone, used apparently as an anvil bone while chipping flint tools on the knee. Bison were then confined to other parts of Europe, so the group had travelled far – a fact further attested by sea shells used in necklaces, and pieces of amber found normally in the Baltic. Elder Bush Cave also had a cache of what had been reindeer meat, buried in a stone-lined box. Penknife blades were found at Dowel Hall Cave in the Upper Dove Valley, with reindeer and bird bones, and an antler point over 11,000 years old. Other caves used by these last British reindeer hunters are One Ash Cave in Lathkill Dale, and Harborough Cave near Brassington, while chance finds have been made at Minninglow, Froggatt Edge, and Totley Moor.

A kind of tundra or moorland landscape, though drier and colder, existed at this time, with dwarf willow, pine, birch, and juniper bushes in sheltered valleys. Cave sites near reliable water supplies were favoured.

Above: Chrome Hill and Hollins Hill with burial mound near the source of the Dove.

Left: Breathtaking view of the Manifold Valley from Thor's Cave

Hunters were based on family groups, and analysis of present day hunter-gatherer societies indicate that composition at campsites would change from day to day, as hunters came and went. Apart from the seasonal following of migratory animals like reindeer over larger distances, the immediate foraging area would have been around a number of more or less permanent water sources.

Overnight or short stay camps might accommodate thirty to fifty individuals. Curiously, it has been established by anthropologists that 150 is about the maximum number for maintaining control of a group by peer pressure alone, and is the basic workable social unit. Most people even today belong to an 'in-group' of twenty-five to thirty friends and acquaintances, roughly approximating to the basic hunter-gatherer band. Three bands would make up a clan, and five clans make a tribe, united by common dialect, and operating in close support geographically. In this respect, our social instincts are still in the Stone Age, while technologically we depend on the microchip instead of the microlith.

MESOLITHIC MISCHIEF

Archaeologist Leslie Armstrong thought he had discovered a female skull, and part of a skeleton dating to the Mesolithic (or Middle Stone Age), at Whaley rock shelter, not far from Creswell, but later digs suggested it was probably a Neolithic burial. The cave, in magnesian limestone, had certainly been used in Mesolithic times, after the end of the ice age 10,000 years ago. The warming up period is now thought to have occurred quite quickly, over a mere ten to fifty years. The reindeer moved away to the cooler conditions they had adapted themselves to, now found in Scandinavia, and the sea levels began to rise again, cutting Britain off with her remnant of ice-age hunters.

The Mesolithic hunter was well-provided with weapons. Warmer, wetter, conditions had meant the re-growth of tree cover, and new breeds of axe and adze heads made their appearance to deal with the abundance of wood as a raw material. Fire seems to have been used, in systematic burnings to create grazing conditions for game animals. Spears and bows were in use, and so was the heavy club. From 9,000 years ago, we have some of the earliest evidence that modern humans attacked and killed one another. A cave in Germany has a pit containing the lacerated bones of men, women, and children, with broken spear tips and arrowheads embedded. Cut marks suggest some were scalped, others decapitated. Similar evidence exists at a cave in Somerset. Other Mesolithic skulls in a Danish museum show evidence that by far the greatest proportion of wounds were inflicted by

Carefully crafted flint spear- and axe-heads found in the Derbyshire uplands.

heavy, club-like blows to the head. A cave painting in Spain shows clearly two groups of archers facing one another and shooting arrows.

Had a golden age of peaceful co-operation come to an end, and if so why? Well, in reality humans had probably been splitting skulls and slitting the throats of one another, and Neanderthals, as well as everything that hopped, jumped, ran, or flew since implements for doing this were developed. From earliest times, there is evidence of cut-marked human bones being de-fleshed and dismembered, even broken to extract the marrow, and the skulls opened. Some are found in ritual grave situations, or just dumped in refuse pits with other food debris. What may have been a family group of two adults, two teenagers, and a three-year-old were found in this condition at Gough's Cave in Somerset. Sadly, modern studies of tribal groups show that overcrowding and competition for resources are not the main causes of conflict. Often, they are based on mere slights, insults, and clan prejudice (or politics!). It is quite probable that increasing numbers of Mesolithic hunters, finding themselves in competition with other groups not in their 'tribe', made opportunistic attacks, like feuding eighteenth-century Red Indian war bands, or indeed seventeenth-century Scottish Highland clans. They may have then eaten parts of their rivals, as cannibals have been doing for millennia. The Gough's Cave evidence shows the remains were processed for butchery in the same way as animal bones alongside, with even the tongues cut out, the scalp removed, and jaws broken to extract marrow. There is no cooking evidence, so it appears the flesh was eaten raw! Human sacrifice by witchdoctors, to purchase favours from spirits, particularly of children, is well known even in modern Africa. In the wild, our closest animal relative, the chimpanzee, often sets

Right: Mesolithic bowmen like these once roamed the hills in a Red Indian existence.

Below: Cut marks on this early modern human jaw from Gough's Cave suggest cannibalism.

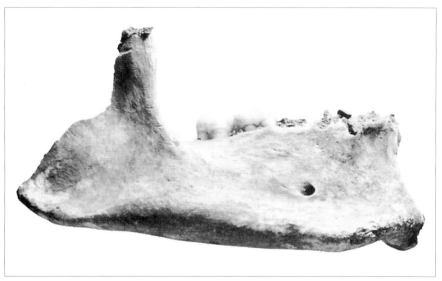

out in all-male raiding parties, simply to attack, torture, and kill members of rival groups. Modern anthropologists now speculate that human and chimp brains are hardwired for this kind of activity.

The light birch forest of the Dark Peak was the happy hunting ground of the Mesolithic people. They used a slender, barbed spear point, made from a strip of deer antler. They also had a broad-leafed flint blade, and lots of tiny bladelets in a variety of shapes, called microliths, which fitted into grooves in the sides of arrow shafts, and were cemented in with tree resin. Broad blades of the period have been found at Fox Hole, in a tall hill near Earl Sterndale in the Upper Dove Valley, with the remains of deer, horse, and wild boar, and a bevelled antler rod. The horse remains show that extensive grassland was still present over the White Peak. Almost anywhere in the Dark Peak moors, where watercourses have cut down to the gravel, and ancient tree stumps stick out of the peat, microliths can be found. Usually of imported flint, or locally quarried black chert, they fell to earth as millions of arrowtips shot, and lost, while the Mesolithic hunters lived a Red Indian existence in the upland woods for some 4-5,000 years.

Thomas Bateman, the author of *Vestiges of the Antiquities of Derbyshire*, and the opener of hundreds of burial mounds in the mid-nineteenth century, has given an excellent description of one of these early hunters, based on his knowledge of artefacts found as grave goods at a slightly later period. He wrote:

> The armillae (worn around the neck or arms) were formed of two boars' tusks securely tied together at each end, so as to form a circle; which would be a very becoming ornament to a hunter, whose dress (probably of deerskin) was pinned together with a bone skewer, and who would be armed with a bow and arrows pointed with flint, having a dagger of the same, or perhaps an adze or hammer, with the head made of stag's horn, suspended from his belt; his limbs covered with red pigment, he would certainly be a formidable looking individual; clothed and armed with articles at once the witness of his ingenuity and prowess in the chase.

Bateman says that women of the period would also have worn skins: 'But we are unable to ascertain the form of the garment.' He asserted that women also wore: 'Suites of beads and armillae of bone or bituminous shale, which when polished appears equal to jet. They also seem to have been initiated into the use of flint weapons.'

The red ochre with which our huntsman may have been adorned is a naturally occurring mineral of clay-based iron oxide, often found placed in barrows as lumps, or sprinkled over the bones of the dead in cave burials.

Fox Hole Cave is near the apex of High Wheeldon Hill, Earl Sterndale.

Bateman uncovered many pieces and wrote: 'The rouge of these unsophisticated huntsmen, even now, on being wetted imparts a bright red colour to the skin, which is by no means easy to discharge.' In addition to his bow and arrows, the hunter probably carried at least one short throwing or stabbing spear, probably not much longer than his bow, for ease of carrying through the tangled forest, and may have been accompanied by a hound, for retrieving or bringing down a wounded prey. This, then, was the type and appearance of our remote predecessor for the greater part of the existence of modern humankind in Britain, whose thoughts, experience, instincts, and aspirations are branded in our natures.

Any vantage point on the gritstone moors usually provides evidence of Mesolithic activity. At Deepcar, near Sheffield, there are concentric rings of stones, which contain masses of flint implements and traces of hearths. Similar sites occur at Totley, Birley Spa, and on Saddleworth Moor, to the north. On Eyam Moor is a rock basin in a large stone, and the marks of some wooden structure probably associated with butchery and the curing of hides. At Wetton Mill Rock Shelter, in the Manifold Valley, were found the remains of flint blades and animals including auroch (a wild ox), and large fish from the nearby river. Dowel Cave revealed a lance made from antler, and patinated blades were found with remains of horse at Calling

Necklace of polished jet, from a burial mound at Grind Low, Over Haddon.

Low Dale rock shelter, near Lathkill Dale. Bones of ox, horse, red deer, and hare were found beside a Mesolithic hearth and anvil stone, in a rock shelter at Stoney Low, near Sheldon. An open air camp here, and at Lismore Fields near Buxton, are two of probably many more sites yet to be discovered in the Derbyshire uplands. Increasingly, people were beginning to build semi-permanent shelters, and cave dwelling became a secondary place of resort. However, the use of caves as homes continued, in some instances, down to the present day.

NEOLITHIC IMMIGRANTS

A revolution in the way people lived began in the Middle East, with the cultivation of crops, and the domestication of farm animals. The Neolithic (New Stone Age) people were the first to practice this, and began to arrive in Britain by sea. Only log boats and skin-covered, wicker-framework boats are known from the period, but later Bronze Age boats were suitable for sea voyages, and it seems sensible to suspect that a sea-going people would have developed something quite similar. So, these immigrants probably arrived sailing their cleverly-constructed plank and skin-covered boats, which were made from split logs bored with holes, and stitched together using sapling strips. Mesolithic people had made canoes from hollowed out logs, but this was real Neolithic naval technology, with sails made from skins, proper wooden keels, and rudders. Anyone who disputes this is welcome to put to sea in a giant coracle carrying a cargo of stone axe heads and see how they get on.

Flint microlith found by the author in gravel near Boggart Stones, a vantage point on the edge of Saddleworth Moor.

Neolithic colonists reached the inland Peak of Derbyshire about 5,000 years ago, bringing grain from their homelands in Turkey (DNA of wheat has been traced there), and started to clear some of the most suitable land for agriculture. At the Bridestones, near Leek, they raised up a permanent symbol of their presence – a Neolithic burial chamber, with an entrance passage made of upright stones covered with a mound. A similar chamber stands on a hill at Five Wells above Taddington, visible for miles around, and there are others on prominent spots in the Peak. Both of these chambers have now lost their mounds, and only the upright stones remain. Because Neolithic farmers settled on the land, and did not roam after wild animals so much as the earlier peoples, this was a statement of their claim to what would become ancestral lands. In effect, it was an artificial cave, a house for the dead. Many more such monuments would spring up all over the Peak in the coming millennia, but for the time being, caves continued to be used as both homes for the living, and sepulchres for the dead.

The Neolithic farmers kept up their hunter-gathering activities, but they preferred the more open grasslands of the White Peak to the wooded Dark Peak, where the earlier Mesolithic peoples continued their nomadic existence, probably regarding the incomers with mounting suspicion, and indeed many Neolithic skulls from graves show indentations inflicted either by violent clubbing or sling stones. They acquired greenstone hammer-axes and maces from Langdale in the Lake District, and others from Graig Llywedd in North Wales, but try chopping down a tree with a blunt stone axe like one of these! Fire was used for extensive clearance.

The dividing line between the open limestone and the wooded moors was along the Hope Valley, and the people on each side of this may have been quite different. Neolithic skulls have long, narrow, boat shapes (appropriate for a sea-going people!) when viewed from above, and their skeletons were of smaller-sized, slender people with refined features. Their skulls are said to generally resemble ancient Egyptians or Sumerians, and they are described as a Mediterranean people. Curiously, tribal boundaries continued to be asserted along this axis even into much later periods – hence the Bronze/Iron Age hillfort on Mam Tor near Castleton, and the defensive earthwork known as the Grey Ditch, at the end of Bradwell Dale, where it joins the Hope Valley. The Romans also built their fort at Brough close by, to guard produce of the lead mines, so tribal boundaries may have continued based on rivalries established in this earlier period, when nomadic hunters may have wished to traverse land which immigrant farmers now claimed as their own.

Oxen, goats, sheep, and pigs were tamed and herded by the Neolithic farmers. The horse had been domesticated by this time on the continent, but there is no evidence for this in the Peak. Wheat, barley, and flax were grown, and flint sickles have been found, along with quern stones for grinding the grain into meal. Flax was used as a fibre for rope, and earthen pots of roughly-baked clay were made. About this time, the first Derbyshire ale was definitely brewed, probably from barley malt. We know this from traces of ale found in drinking vessels buried with the dead.

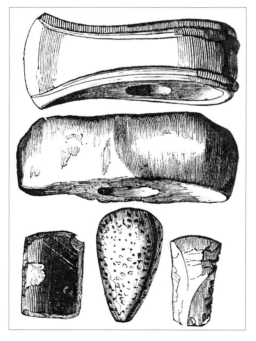

Stone axe-heads recovered by Bateman from Derbyshire burial mounds.

Caves continued to be used for habitation, and also for burial. At Calling Low Dale, off Lathkill Dale, an area of overhanging cliffs forms a dry, rock shelter, where a collective burial was made of sixteen individuals. In a cist against the rock face were found two skeletons, with shards of a broken bowl and a flint arrowhead. Seven Ways Cave, on Thor's Cliff, produced skeletal remains, and two leaf-shaped arrowheads. Another arrowhead was deposited with human bones at Dafar Ridge Cave, Manifold Valley, and Rains Cave near Brassington held human remains, kite-shaped arrowheads, and two Neolithic pots. At Dowel Cave, ten skeletons and the headless remains of a dog were found. Other Neolithic remains were found at Megdale, Matlock, and in Fox Hole Cave, High Wheeldon, where a polished stone axe and bowls were recovered. Caves above the Lower Manifold Valley at Falcon Low and Cheshire Wood also produced human remains and pottery. Harborough Cave near Brassington and Ravencliffe Cave, which had been in use since Old Stone Age times, also held Neolithic evidence.

It seems that the Neolithic immigrants in the Peak never represented a major population, and were probably assimilated by the aboriginal inhabitants to a large extent. The next cultural wave was from Europe at the beginning of the Bronze Age, with the coming of the Beaker Folk nearly 4,000 years ago.

Quern stone found on Harthill Moor, near remains of a Neolithic settlement.

WIZARDS OF METAL

With one mighty bound, mankind was freed from the constraints imposed by working with basic wood and stone. This was the beginning of real technology – creation of metal tools, horse harness, woven cloth, the wheel – and the process of transformation was a kind of magic!

Stone quarried from the earth, when fired in a forge, produced a mysterious molten liquid – which could be guided into a clay mould to make axes, swords, daggers, and tools. This first smelting of copper, from the green or yellow mineral early man may have puzzled over as it disintegrated, leaving a shiny metallic residue in the ashes of yesterday's fire, provided the stepping stone to the new world.

Remains of the Beaker people – so called after their distinctive, beaker-shaped cups, found in the round barrows where they buried their dead – show a taller, strongly-built racial type, with more rounded skulls and bigger jaws than the Neolithic farmers. They brought implements of bronze and brass, but flint continued in everyday use, and is often found in burial mounds. At Fox Hole Cave, Beaker folk laid a cobbled floor through ninety feet of passageway. They also left the remains there of several brown bears they had killed and eaten, besides their domestic

Seven Ways Cave on Thor's Cliff top.

animals, and red deer, roe deer, and birds. As boar's tusks and bear's teeth are known to have been prized necklace ornaments, the cave may well have had the status of a hunting lodge, with its excellent viewpoint high on a hillside above a narrow valley. At Wetton Mill rock shelter, they left beautifully tanged flint arrowheads, alongside beaker-type pottery, bone pins, and spatulas. A bear's tooth pierced for a necklace was found at Harborough Cave, and boar's tusks have often been buried in Bronze Age burial mounds in the Peak, frequently raised over natural fissures in the rock, in which the remains are deposited. The boar continued to have a special place in hunting mythology, venerated for its fierce courage and intelligence in the chase.

Later Bronze Age remains found in caves included amber beads, and a lovely, polished jet button, with the bones of a dolphin, buried in Thor's Fissure Cave – evidently of great ceremonial and mystical value so far from the sea. Again at Harborough Cave was a bronze knife blade and two gold rings, and at Ravencliffe Cave finds of pleated gold bands show caves continued to have ritual significance. Ireland was the main centre of bronze working, and most gold was also imported from there, but itinerant bronze smiths, regarded as having almost wizard-like powers, travelled about, smelting, and mending broken weapons and implements. Tin obtained in Cornwall was needed to mix with copper to make bronze, so journeying for trade became a necessity. The outcroppings of copper ore at Alderley Edge had been worked with antler picks from the beginning of the bronze period, and became sizeable mine workings by the late Bronze Age. Copper deposits beside the Manifold Valley at Ecton Hill were also exploited, but these lay in harder rock and deeper underground – not fully developed until the eighteenth century when they formed, briefly, the largest copper mines in Europe. But an antler pick from here has been dated at 3,700 years old, proving they were worked even long before the Romans. The story of mining in the area from earliest times can be seen, with working exhibits, at the Peak District Mining Museum, Matlock Bath, for a modest entrance fee. Temple Mine, behind the museum and the old Pavilion, provides an easy descent through horizontal tunnels into the hillsides. They also run a field centre at Magpie Mine, near Sheldon, where serious visitors can descend a vertical shaft into an old lead mine.

Towards the end of the Bronze Age and beginning of the Iron Age, the weather became cooler and wetter, making the land more difficult to farm. A series of hill-forts seem to have been erected in a stupendous civil engineering project, on an axis running east/west across the Derbyshire Peak. Beginnings of this are found at a banked enclosure at Gardom's Edge, on the eastern moors above the Derwent, continuing via

Above: Nan Tor rock, above the bridge over the Manifold at Wetton Mill.

Left: Crude early earthenware food vessels unearthed at Arbor Low stone circle.

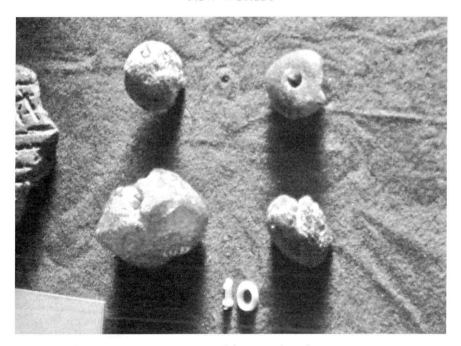

Amber beads at Buxton Museum – prized for special qualities.

Ball Cross at Bakewell, to Fin Cop above Monsal Dale, via the gritstone walls of Carl Wark, on moors near Hathersage, to Burr Mount above Great Hucklow, and beyond mountainous Mam Tor, towering above the Hope Valley, to Castle Naze on Combs Moss and Mellor hill-fort in the west. Anyone who has examined the ditch and bank system around the summit of Mam Tor, or at Castle Naze, will realise that this was a major undertaking, and reflects an important tribal boundary intended to be guarded at all costs. JCBs were then unheard of, and all this soil and rock had to be shifted by hand. It has been suggested that the threat came from the chariot-riding Belgae – another continental incomer, who may have had an eye on the rich mineral pickings of the area, but at all events this provided a nice dividing line between the tribes of the Coritani, allies of the Belgae, to the south, and the powerful confederation of northern tribes, known later to the Romans as the Brigantes, whose bright red hair was a powerful signification of their link via Mesolithic ancestors of the forested uplands, to the DNA of the hunter gathers of the last ice age.

Derbyshire caves continued in use throughout the Iron Age, and during Romano-British times, and there is evidence for a bronze smith at work in the entrance to Poole's Cavern, Buxton, in the second century of the Roman occupation. Caves were used as sepulchres even by the pagan Saxons, who buried a boar's crested helmet in a barrow at Benty Grange.

Mam Tor hill-fort, from Rushup Edge above Edale.

Thirst House cave, Deepdale, near Buxton – used by outlaws, Romans, and early man.

In the medieval period, local caves became bases for outlaws in the royal hunting forest, and later gypsies and tinkers made use of their rent-free facilities, with one old woman in the nineteenth century, called Straw Legs, notorious for abusing people in the neighbourhood of Buxton who would not buy her trinkets. She lived in Thirst House Cave, Deepdale, and bound up her legs with straw to keep warm.

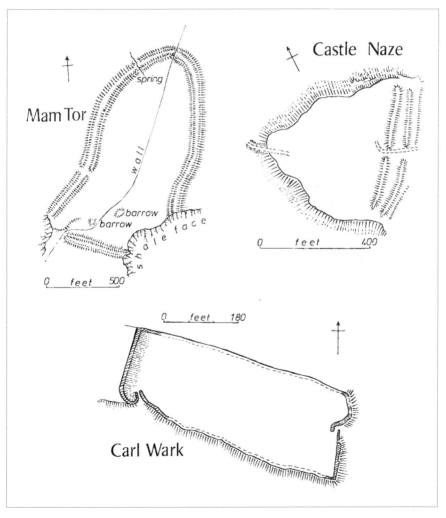

Iron Age hill forts.

3

WONDERS OF THE PEAK

TALES AND LEGENDS

Among the seven wonders of the Peak in earlier times were listed three caves – Eldon Hole, believed to reach down into Hades, Peak Cavern (or the Devil's Arse), and Poole's Cavern, refuge of a robber. The Norman historian Henry of Huntingdon called Peak Cavern the first wonder of England. A thousand years ago, it was believed that: 'A gale leaves the caverns in the mountain called Peak with such force that clothing cast into its path is blown high up and jettisoned far away.'

An early thirteenth-century romance described a swineherd wandering into Peak Cavern in search of his lord's favourite sow, and finding her in an enchanted, underground, sunlit land of cultivated fields, happily grazing! Less pleasant associations are with the Devil dining there with gypsies and robbers, and producing stinking blasts from his digestive process (hence the gale!). Poole's Cavern at Buxton is also linked with a notorious thief called Poole, a sort of Robin Hood, and Eldon Hole was visited by Queen Elizabeth's favourite, the Earl of Leicester, who bribed a servant to be lowered in a basket to the rope's limit of 200 yards, and was brought up a nervous wreck, unable to speak. He died in a demented frenzy eight days later, so the Earl decided the unfortunate man must really have glimpsed Hell! Other tales tell of a goose thrown down Eldon Hole emerging from Peak Cavern, two miles away, with its feathers singed off by the fires of Hell. An old ballad, 'The driving of the deer' calls it Hellden Hole.

Eldon Hole is the only naturally open, true pothole in the Peak, according to Derbyshire's eminent cave expert, Dr Trevor Ford, and was, for many years, believed to be of unfathomable depth. Doubtless, it was known by early man, but whether he was able to enter it and leave any evidence, other than offerings flung down the shaft is difficult to ascertain. There are later tales of robbers throwing their victims down

Impressive gorge created by Peakshole Water leads to the entrance of Peak Cavern at Castleton.

its impenetrable deeps. The 100 foot cleft lies on Eldon Hill, described as the best-known eyesore in the Peak, though it doesn't look so bad now weathering has mellowed the rock face since the quarry was abandoned in recent years. It is on the furthest northern exposure of the limestone before it gives way to the grits and shales of Rushup Edge and the Kinder plateau. On its summit is a Bronze Age burial mound, and on its southern flank lies Eldon Hole, facing towards the little village of Peak Forest, once the medieval administrative centre of the royal hunting forest.

Curiously, in 1250 it was named as Elvedon Hill, or the *hill fortress of the elves*, and was believed by the local people to be populated by elves, like many of the caves in the locality. In fact, elves, or 'the knockers', were often believed to mislead, or sometimes help miners, and were propitiated with offerings of food down to recent centuries. William Camden visited Eldon Hole while compiling a gazetteer of Britain in Elizabethan times, as did the angling squire, Charles Cotton, in about 1680. Cotton claims that 800 fathoms of line were let down, and still the bottom was not plumbed. The journalist and author, Daniel Defoe, was impressed by this tale on his visit, concluding: 'Perhaps this vast chasm goes directly down to the centre of the earth!'

Eldon Hill from Perryfoot in the snow – no elves in sight, just an overdressed author.

Disused in recent years, the unstable ledges at Eldon Hill quarry exposed an unknown cave system possibly linking with Eldon Hole.

But in 1761, a man called Lloyd was lowered into the hole, and reached the rubble-covered floor of the first cave within about 200 feet, where there was still enough light 'to read a book'. He then explored another passage leading to a great cavern with a further long drop in its floor to another cave with a river flowing through it. He published his findings for the Royal Society in 1780. There is now no trace of the further drop to the cave with a river, and in fact, so many tons of rubble have been hurled down by vandals, who have demolished successive stone walls, built to surround the chasm, in order to test the depth, that the actual cave floor is completely buried. Perhaps one day, when stone removal and digging techniques have sufficiently improved, the bottom may be fully excavated and ancient animal, and, perhaps, human remains may be revealed.

A fine description of this rock-strewn bottom was given by Ernest Baker, of the Kyndwr Club, when he descended in 1900.

> The bottom is an irregular oblong in shape and the sloping floor is covered with broken rocks. It is a grim and gloomy spot lit by a very small patch of sky. All view of my comrades was cut off by overhanging ledges. On

Swarthy and hirsute members of the Kyndwr Club prepare to descend Eldon Hole by bosun's chair in 1900.

Another view of Deepdale from the cave's mouth.

one side the floor falls away rapidly to the mouth of a small cave, which I resolved to explore. Crawling under the portal I found myself on a slope covered with stones of all sizes in a very unstable condition.

With better lighting than his single candle, Baker was eventually able to discern the vast cavern he had emerged into, with a white arch 'exquisitely symmetrical', and high above, a great dome with creamy white and reddish brown stalactites, 'recalling the chiselled walls and vaulting of a cathedral.' But of the further passage in the floor described by Lloyd, he could find only a blocked hole. It is now known from dye tests that water drains eventually into Russet Well, Castleton.

Another locality cited by the Peakrils (inhabitants of the Peak) as prevalent with elves is Deepdale, a limestone gorge with a seasonal stream in the bottom, dry in summer, flowing in winter, with several caves, not many miles from Buxton. There is an old poem which runs: '*When in Deepdale wear not green, lest you offend the Faery Queen.*' John Ward, a Fellow of the Society of Antiquaries, wrote that Thirst House Cave, generally held to be haunted and the abode of fairies, sometimes confused with Deepdale Cavern, derived etymologically from Hob's House, hob being a tricksy elf, and via Hob Hurst House, devolving to Th'Hurst House. Hurst being a capricious wood elf, and the name Thirst House not referring to the lack of water in the cave, or in the valley bottom during summer, as a pot strategically placed beneath drips in a Peak District limestone cave will gather water at most times of the year. He told the story of a farm labourer or miner, whose journey to work passed by Deepdale early one morning, where he encountered a small creature dressed in green; seizing it, he put it in a bag and carried on his way. 'But it shrieked piteously until he grew tired and released it, whereupon he watched it as it ran back across the fields to Deepdale.' It is said that Thirst House once had a chamber with a stream, but this is now lost. Deepdale Cavern, across the valley, is a steep winding cave, and there is another small cave beside Thirst House, alongside the path up to Chelmorton. At Churn Holes is another hidden cave, once inhabited by Romano-British people. In more recent times, a lorry driver who slept in his cab at a nearby limestone quarry awoke early, to find 'a group of strange green hooded beings capering about, who then did a kind of slow dance in a circle and vanished.'

Long before Straw Legs in the 1870s, Deepdale caves are said to have been inhabited by medieval outlaws, who flouted the game laws and robbed travellers as they saw fit in the hunting forest. The stories of elves probably kept the locals away, and no doubt the King's foresters would have some kind of working arrangement with the outlaws. In Victorian times, these caves were excavated extensively by a Buxton shopkeeper and antiquary, W. H. Salt, who published his findings, *Ancient Remains Near Buxton*, in 1899. He found many Roman and other objects, and near the entrance to Thirst House, in a shallow, stone-lined grave, were the remains of a tall man with a spear. Local legend maintains that the valley was the ancestral burial ground of the Brigantian chiefs. In Celtic legend, fairies have always been associated with the Druids – lawgiver priests of the original inhabitants of Britain. Merlin was said to have been a Druid, and of course, according to legend, used his magic to work many mysteries. Perhaps the tales of elves are a remnant of the time when such sprites guarded the burial places of the tribe.

The name Peakril, which was applied to the locals by early travellers through the area, seems to be derived from *Pec*, which appears in the Tribal Hideage of the seventh century, a Saxon document where *Pecsaetna lond* means the territory of the Peak settlers. The *Anglo-Saxon Chronicle* of 924 describes Bakewell as in Peakland, which comprised the areas around Buxton, Bakewell, and Castleton, later referred to by 1196 as the High Peak. The Peakrils themselves were not always highly thought of by visitors. Defoe described them as a 'rude, boorish, sort of people,' and later visitors were shocked by their hard drinking; even the women's behaviour was lewd and lurid. Seventy years after Defoe's visit, some Fellows of a

Cambridge College touring Derbyshire encountered drunken women harvesters, 'swearing and using every indecent term they could think of...a bacchanalian revel in the broad face of day.' Harvesting is thirsty work, and in their defence it must be said that people in lead mining districts were forced to drink beer, and advised to do so by doctors, because local water was often contaminated with toxins, and could lead to Derbyshire neck – an enlarged thyroid gland and other ailments. There is an old saying – *'Derbyshire born an' Derbyshire bred, strong in th'arm and weak in y'ead* ' – a covert reference to lead poisoning. A traveller through Taddington in 1811 complained it was 'a dirty, cheerless, inhospitable village, inhabited by a race of Peakrils as uncouth and barbarous as a race of savages.' Their only virtue was a tendency to longevity, given 'the healthfulness of its situation and its good air.'

Peakrils were the last cave dwellers in England, inhabiting the very places where they earned a living as lead miners or lime burners. Miners and their families lived in caves close by the mines, and mounds of lime were made into dwellings by the lime burners, so that Grin Low above Buxton was festooned with these low dwellings, 'appearing like an assemblage of tents' one above another, some with cows on the roof, shrouded in the smoke from the lime burning. A clergyman visiting Poole's Cavern in 1800 described being confronted by nine old women, 'dried with age and as rugged as the rocks amongst which they dwell', who conducted him on

This poor old woman who loved a pipe and a jug kept the well at Buxton, and lived to be 90.

a tour of the cavern, this right having been granted them by the Duke of Devonshire, their landlord.

An entire village of rope makers inhabited the entrance to Peak Cavern up until about 1845, and had done so for hundreds of years. The streets for the low, huddled, thatched cottages form several terraces just inside the cavern, and soot from their fires still blackens the roof. The appearance of the Peakrils was often unedifying. The women wore trousers and caps, and chewed tobacco; their dialect was difficult to understand for outsiders, and the miners were pale and dirty through being continually underground, with a sickly yellow tinge caused by breathing the poisonous fumes from the lead smelting, and handling the ore. The expression 'looking peaky', meaning unwell, is still used in the North of England. By the seventeenth century, there were reckoned to be 4,000 free lead miners of the Peak working on their own account, with a further 21,000 dependent on the industry.

Daniel Defoe visited an interesting cave at Harborough Rocks, near Brassington, where he met a lead miner's wife in about 1720. The cave, which still exhibits some of the features he mentions, has been in use since at least Late Palaeolithic times. She appeared at the entrance, with a babe in her arms, and a little girl at her side, while their small dog ran about barking.

Harborough Cave, where ancient people lived for millennia, and Daniel Defoe met a lead miner's wife and her children.

Good wife, where do you live? "Here, sir," says she, pointing to the hole in the rock... And do all these children live here too? "Yes sir, they were all born here, and my husband and his father before him."

Defoe was invited inside.

There was a large hollow cave, which the poor people by hanging curtains across had partitioned into three rooms. On one side was the chimney and the man or his father being miners had found means to work a shaft to carry the smoke out at the top. It was poor, but not as miserable as I had expected. Everything was clean and neat, though mean and ordinary. There were shelves with earthenware, and some pewter and brass. There was, I observed, a whole flitch or side of bacon hanging up in the chimney and by it a good piece of another. There was a sow and pigs running about at the door and a little lean cow feeding upon a green place just before the door and a little enclosed piece of ground growing with good barley then near harvest.

Defoe, with this description, gives us some idea of how life in a cave may have been rendered habitable and, leaving aside the pewter

The view from Harborough Cave takes in some industrial scenery as well as pastoral.

and brass, perhaps not so unlike the lives of the countless generations stretching back over the millennia who had inhabited the place. The floor of Harborough Cave is now a little uneven, having been repeatedly excavated, but it still contains the chimney and the little antechambers, which with a slight effort of imagination, can be transformed into the dwelling of this eighteenth-century family of troglodytes.

On learning that the family worked hard, yet lived on a mere eight pence a day, Defoe and his friends clubbed together to give her a present amounting to about five shillings. On presentation of this, she burst into tears. 'All she could do for some time was weep, then she thanked us as handsomely as a poor body was able,' she said she had not seen so much money together in her hand for a long time. Asked if she had a good husband she replied, 'Yes, thank God,' that apart from being short of money they were very happy, and lived contentedly. 'This made our party very grave for the rest of the day,' commented Defoe, 'that people could be content in such poverty. The woman was tall, well shaped, and clean, nor was there anything like the dirt and nastiness of the miserable cottages of the poor, though many of them spend more money in strong drink than this poor woman had to maintain five children with.'

Harborough rocks, above the High Peak Trail, are dolomitised limestone, which is supposed to account for their broken appearance because of fracture-forming minerals.

Above Harborough Rocks, at the nearby trig point, are the flattened remains of what was a megalithic burial chamber dating from Neolithic times, where the bones of five people, and three leaf arrowheads, were recovered from a natural cleft in the rock within the chamber when it was excavated in 1889. The country people said it was the tomb of a giant, which had a slab about 16 feet long upon it at one time. A local farmer told Ernest Baker, author of *Moors Crags and Caves*, who visited in 1900, that the cave had Druid associations, and pointing to the outline of a crucifix on a wall, claimed this as evidence that a hermit may have used it in medieval times. A 'Druid's Chair' is also situated on the rocks close by remains of the burial mound. This is a lump of limestone, fashioned into the likeness of a bardic throne, and has recently been damaged by some witless vandal who chipped a piece off one of the arms. It is in a grand situation, facing a natural amphitheatre – well calculated to hold an awestruck audience in thrall for a magic show of celestial and meteorological wonders – but accurately dating megaliths can be difficult. It may just have been an eighteenth-century folly carved from a natural stone.

A strange tale in an old book describes a rock shelter, which produced Iron Age relics now held in the British Museum. It is called Old Woman's Cave, and is situated above Taddington Dale, near the start of a small

This limestone block had an eighteenth-century inscription, and is said to be a Druid's chair. There was a nearby Megalithic burial chamber.

Demon's Dale Cave, an old resurgence cave above the footpath leading from the car park at the bottom of Taddington Dale.

valley once known as Demons Dale, and mentioned as such in old documents. This vicinity is rich in ancient remains. On the hillside nearby was an area used as a seasonal camp from very early times by hunters. A number of valleys meet at this point, and the River Wye, which may have had religious significance for early peoples, flowing as it does from a vast underground cavern in Buxton near hot springs, glides down Monsal Dale, below the fortress of Fin Cop, and past Great Shacklow Wood, towards Ashford and Bakewell.

William of Worcester, writing in medieval times, recorded: 'The Wye runs by the town of Monsal and a valley called Demonsdale, where spirits are tortured, which is a marvellous entrance into the land of Peke, where souls are tormented.' Local legend tells of a spirit trapped beneath a flat stone in the river opposite Demons Dale, not so far from the modern car park built for walkers and those come to admire the beauty of the Peak. A recent dig uncovered a skeleton flung into a ditch, which surrounds the outer defences of Fin Cop on the steep-sided hill opposite, and another was previously exhumed from a burial mound at the hill-fort, where a male skeleton was buried face downwards, with a slab of black marble covering his skull. Placed alongside were two flint arrowheads and a small flat circular stone, coated with yellow stucco and varnished.

Hazelbadge Hall, the ancient home of Margaret Vernon, broken-hearted lover turned ghost rider!

Looks like the crater of an extinct volcano. Cowhole Cave, viewed on cliffs above Bradwell Dale.

Farther up Monsal Dale is Hob's House Cave, which a gigantic being was said to haunt. The skeleton of a young person was found here with Bronze Age pottery, while at Demons Dale Cave the remains of four people were uncovered, along with a flint dagger and pottery dating from Neolithic to Bronze Age times. No-one can now explain the meaning of a local rhyme:

> *The Piper of Shacklow,*
> *The Fiddler of Fin,*
> *The Old Woman of Demon's Dale,*
> *Gather them all in.*

Finally, there is the romantic, but tragic, haunting cave story of Margaret Vernon of Hazelbadge Hall, Bradwell Dale, a young woman who, centuries ago, fell in love with a handsome man who did not reciprocate her love, but married another. Margaret rode from her ancient, grey, brooding home, built in 1549, through the narrow defile of the dale to Hope Church to see him marry another woman. She galloped home, driven crazy by her unhappiness, and on wild nights, it is said she can *still* be seen riding hell for leather on a white horse past 'the beautiful little cavern, shining with stalactites,' which is mentioned as the Great Cave of Hazelbadge in a document marking the bounds of the King's Forest of the Peak. However, on a sober note, I've looked for this, and it seems to be on the cliffs, about 100 feet above the road, so she really must be riding a ghostly steed. Unless, of course, the writer simply confused it with the now-overgrown Bradwell Dale Cave in the valley at the entrance to the village. Many cave names get confused over time.

4

ANTIQUARIES AND BONE CAVES

HERMITS V. DRUIDS

The early antiquarian traveller, William Stukeley, first created an interest in the ancient curiosities of the area after he recorded two visits in his *Itinerary* of 1724. The first cave he found was a hermit's cell, carved from a rock beside the River Derwent, among the ruins of Dale Abbey, near Derby. Intrigued by this contemplative life, he next visited the Hermit's Cave at Cratcliff Tor, near Birchover, where he made a sketch of the carved crucifixion. 'I entered another hermit's cell, who had a mind if possible to get quite out of the world; 'tis hewn in the rock with a most dreary prospect before it. On one end is a crucifix and a little niche where I suppose the mistaken zeal of the starved anchorite placed his saint or such trinket.'

The situation now is delightfully shaded by mature yew trees at the foot of the rocks, with an iron railing erected in the nineteenth century protecting the cave, to a certain extent, from the attentions of rock climbers who started their pilgrimages to this place in the 1890s to test the skills of their fledgling sport.

The cave overlooks the ancient Portway, a major route through the Peak since Bronze Age times, now a steep trackway, and travellers would often visit such anchorites' cells to ask a blessing on their journey, and leave some offering for the holy guardian of the place. The Portway passes between here and the strange, chimneyed pile of gritstone rock known as Robin Hood's Stride, not far off, and a stone circle on Harthill Moor, where a herd of wild fallow deer still roam. Recently, archaeologists have found evidence of a defensive structure on top of Cratcliff Tor, probably dating from at least the Iron Age, and there is another, named on the map as Castle Ring, surrounded by ditches and defensive stonework, half a mile further on. Neither has been properly archaeologically investigated.

Stukeley crossed the valley to look at the boulder-strewn hillside of Rowtor Rocks, where a local vicar had made a rock-garden of hanging

Medieval crucifixion scene and niche for lamp, carved by an anchorite (solitary) in the Hermit's Cave at Cratcliffe Tor near Birchover.

Victorian tourists flocked to see the hanging gardens, and romantic rocks and grottos with Druidic associations at Rowtor Rocks, near Birchover, where a cottage became the Druids Inn.

terraces. 'I was highly pleased with so elegant a composure, where art and industry had so well played its part against rugged nature,' he recorded. The alcoves and caves gave him an idea to create a 'hermitage' in his own garden at Stamford, soon imitated in fashionable parks and gardens by the gentry, who even paid 'hermits' to hang about in them, startling their guests with demands for 'alms', and adding to the wild, gothic taste. Another antiquary, visiting sixty years after Stukeley, Major Hayman Rooke was convinced that this was a genuine Druidical Grove, with rocking stones, and rock basins for holding the blood of sacrificial rituals. He described the Nine Ladies stone circle on nearby Stanton Moor as a Druidical Temple, and that on Harthill Moor also. But this was by the time that Stukeley himself, having written about Avebury and Stonehenge, had created a firm belief in Druidism as the ancient, patriarchal religion of Britain. Coach-loads of tourists began to arrive, the rocks were fenced in, and a keeper was established in a cottage at their foot, soon enlarged to become an inn named, naturally, 'The Druids.'

BARROW KNIGHTS

Thomas Bateman, the 'Barrow Knight', was both lauded and derided for his grave-opening activities in early nineteenth-century Derbyshire. A rich squire of Middleton-by-Youlgreave, he had inherited his first fortune from his father at the early age of twenty-one, and built himself Lomberdale Hall in the village, where he housed his burgeoning collection of antiquities, purchased from farmers and collectors, and taken from caves and burial mounds throughout the Peak. The local newspapers denounced his most controversial act, that of taking a hairpin from the skull of Dorothy Vernon, when her tomb was opened during the demolition and rebuilding of Bakewell Parish Church. Dorothy was the famous heiress who had eloped with a lover from Haddon Hall in Elizabethan times. It is said her skeleton collapsed in a pile of dust after it was so rudely disturbed.

Bateman's early views on religion are summarised in his *Vestiges of the Antiquities of Derbyshire* thus:

> All people, both ancient and modern, however barbarous and uncivilised, have an innate foreboding of the future, and a consciousness of the existence of an omnipotent power, which would influence the future, which they desire to propitiate, and which they cannot think of without a feeling of reverential awe. These ideas have inevitably led to the establishment of some form of worship, the erection of places to worship in, and the assumption of sacerdotal office by some of the more cunning or better informed members of the tribe or nation.

Thomas Bateman amassed a unique collection of antiquities, which he had wrested from the earth of Derbyshire and Staffordshire.

Young Thomas had laid the foundation stone for his grandfather's Congregational Chapel in the village of Middleton at the age of four, but a religious spirit did not seem to influence his later actions as a young man. His grandfather was scandalised when his grandson went to live with a married woman in Bakewell on inheriting his father's fortune. As soon as Lomberdale Hall was completed, Thomas moved in with this woman and her younger sister, creating rumours of a *ménage a trois*! Thomas had also inherited a taste for practical antiquarian research from his father, and set about opening yet more of the burial mounds with which the Derbyshire and Staffordshire hills are abundant, employing his fortune to both bribe landowners and pay a gang of labourers, who, on occasion, 'did' as many as three burial mounds and their contents in one day. So it was that most of the ancient chieftains of the Peak were rapidly disinterred in a few short years, and denuded of their grave offerings after lying in peaceful, if not reverential, state for upwards of five millennia. He is said to have opened about 300 barrows between 1847 and 1851, before illness cut short his career.

In 1847, his grandfather died, and a shocking stipulation in his will was that Thomas should 'end his criminal association' with the married woman, or he would not inherit Middleton Hall and his grandfather's vast estate of 3,000 acres. Ever the pragmatist, Thomas disentangled himself, and within two months had married Sarah Parker, the housekeeper at Lomberdale Hall. He was now the owner of two halls, and the village of Middleton. The library and museum at Lomberdale were extended to become one of the greatest collections of early antiquities in Britain.

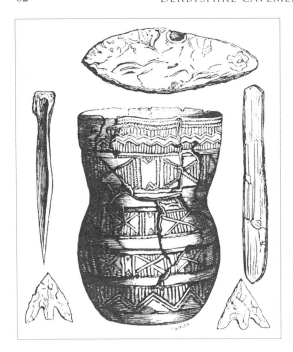

A decorated drinking cup, bone pin, spatula, flint arrowheads, and a flint dagger blade, found near the contracted skeleton of a man by Bateman at Green Low near Biggin.

Although his methods of excavation were crude, he was a detailed note taker, and on the publication of *Vestiges* was already a member of the British Archaeological Association.

Illness in his thirties drove Thomas back to religion, and he and his family started to attend a Sunday school in the basement of his grandfather's chapel in Middleton. By 1853, he had used some of his money to build another Congregational Chapel at nearby Youlgreave. But death came suddenly, during dinner at Lomberdale, from a massive haemorrhage at the early age of thirty-nine, in 1861. His wife did not long survive him, and the pair now lie together, in a curious tomb, surrounded by iron railings, and surmounted by the stone replica of a Bronze Age cinerary urn, in a field behind what was Middleton's Congregational Chapel, but is now a private house. Their only son, a heavy drinker, died insolvent, having sold his father's collection of antiquities, most of which can now be seen at Sheffield's Weston Park Museum.

Thomas Bateman, like his father William, was a pioneering archaeologist, and recorded and saved many finds previous to his own excavations, by labourers, farmers, and curious clergymen, which might otherwise have been forgotten and lost. His greatest discovery was, perhaps, the Anglo-Saxon helmet, with its gold boar's crest set with garnet eyes, from a barrow at Benty Grange, not far from the famous henge at Arbor Low; warrior apparel of an early chief of the *Pecsaetna*, of whom only his hair

Bateman's tomb, and that of his young wife, rests in a field behind the old chapel his grandfather had endowed, now a private house.

remained. Bateman frequently records the numerous rats' bones in ancient burials, and concluded that 'these cannibals' burrowed into the burial mounds, and feasted on the contents, sometimes dragging parts of the skeletons into unlikely crevices. He doubtless made provision against this eventuality in his own funeral arrangements, having hewed the tomb from solid limestone, and leaving instructions for it to be adequately sealed to thwart the attentions of tomb raiders, both animal and human!

BONE CAVES

Throughout his long and successful career, Sir William Boyd Dawkins was mostly known to the public as an archaeologist, and an expert on ancient mammal remains, painstakingly recovered from caves throughout Britain, but particularly from Windy Knoll fissure, near Castleton. Bison bones from here have recently been dated to 37,300 years old. But, in fact, he was Professor of Geology at Owens College, the forerunner of Manchester University, and carried out important work in the first Geological Survey of Great Britain, was surveyor for the first Channel Tunnel Committee, and discovered the Kent coalfield. It was for his services to geology that he was knighted in 1919.

Born the son of the vicar of Buttington, near Welshpool, in 1837, he had started to collect fossils at the age of five. After graduating from Oxford with a first in natural sciences, and working with the Geological Survey, he was appointed curator of the Manchester Museum in 1869, and went to brush up on classics with the Rev. J. Williamson, vicar of Wookey, in Somerset. The two men became involved in an excavation at a large natural cave, Wookey Hole, where they discovered man-made flint instruments, with the bones of mammoth, hyena, and woolly rhinos which 'proved their contemporaneity with extinct mammalia' as Boyd Dawkins put it. Much later, he was able to identify the first remains of a sabre tooth cat found in Britain, at Dove Holes, near Buxton.

His book *Cave Hunting*, published in 1874, established his reputation with a wider public as a foremost expert on the history of early man and extinct, ice-age animals. Having been invited to take part in the investigations at Creswell caves in 1876, he made the astounding discovery of the most celebrated piece of ancient art found in Britain – an engraving of a wild horse on a rib bone from Robin Hood Cave. Recent research by the British Museum has detected red ochre pigmentation, and dated the artwork to around 12,500 years old, near the end of the last ice age. However, because of some jealousy among the excavators, he was accused by one of them, Thomas Heath, curator at Derby Museum, of having planted this piece in the dig! This was mainly because Heath felt snubbed when the excavation report had failed to mention him, and Dawkins had taken the best finds for exhibition at Manchester, and got most of the publicity.

A further book by Dawkins, *Early Man in Britain*, in 1880, summarised his extensive practical research. He had tried to obtain the chair of geology at Cambridge University, and was supported by a testimonial from Charles

Sir William Boyd Dawkins as a young field geologist and explorer of antiquities.

Darwin, but was unsuccessful. Instead, he obtained the chair of geology at Manchester University, to which he was appointed in 1872, a post he retained until he retired in 1908. His style of writing was florid, and full of descriptive High Victorian prose. He correctly attributed remains in a cave in Denbighshire to 'an Iberian or Basque race', which he termed *Brit Welsh*, which he assigned to the Neolithic period. Like many contemporaries, he was keen on measuring the heads of native people to establish their descent, and carried out these investigations on the inhabitants of the Isle of Man. Eugenics was then considered a science, and an undisguised Victorian admiration for the Anglo-Saxon invaders of post-Roman Britain echoes in his sad depictions of Romano-British refugees fleeing to the safety of caves before their unstoppable advance. He quotes Gildas, a witness of this time, describing the catastrophe:

> *The fire of vengeance did not cease, until, destroying the neighbouring towns and lands, it reached the other side of the island and dipped its red and savage tongue in the western ocean... In the midst of streets lay the tops of lofty towers tumbled to the ground, stones of high walls, holy altars, fragments of human bodies covered with livid clots of coagulated blood, looking as if they had been squeezed together in a press and with no chance of being buried, save in the ruins of the houses, or in the ravening bellies of wild beasts and birds.*

A truly shocking scene – but now it appears from DNA evidence that the effects of this, though terrible at the time, had considerably less impact on the gene pool than Victorian historians had imagined.

Sir William Boyd Dawkins was the founder, and first President, of the Lancashire and Cheshire Antiquarian Society. As a philanthropist, he fought to improve the conditions and opportunities for learning of the ordinary worker, and gave generously to various public projects. On his death in 1929, his entire library, and most of his papers, including research notes, and personal possessions, were donated at his request to Buxton Museum, where many are on permanent display, in a room laid out as the archetypal Victorian bone-cave specialist's study. To modern eyes, it looks dark and rather bizarre, with its skulls and specimens – including a stuffed dotterel from Darwin – but here, and in rooms of this kind, men with inquiring minds and little to go on but their own researches and interpretation of finds, laid the foundations of understanding our place in the world – where we came from, and where, possibly, we might be going!

His associate, the author of *Barrows and Bone caves of Derbyshire*, was a Bolton solicitor and amateur archaeologist, J. Rooke Pennington, who, in the 1870s, co-operated with Prof. Boyd Dawkins in the excavations

Castleton had a museum of local antiquities and minerals, founded by Bolton solicitor Rooke Pennington.

at Windy Knoll, near Castleton, which recovered thousands of ancient animal bones. He was a keen opener of burial mounds, and managed to find human remains in barrows already investigated in a tearing hurry by Thomas Bateman. His self-financed museum at Castleton was full of interesting geological and antiquarian specimens.

Pennington and Boyd Dawkins, assisted by John Tym, a former spar worker, and leasee of the Speedwell Mine, supervised a gang of miners, who sifted through the clayey deposits at Windy Knoll, where hundreds of animals had fallen and become trapped in a hidden fissure at the bottom of a muddy waterhole. They found the remains of ice-age mammals, including bison, reindeer, bear, wolves, and a big cat, probably related to the sabre tooth tiger.

Tym became the curator of Pennington's museum in Castleton, where many of their finds were displayed, and he also built a magnificent Blue John window from the local spar, for which Castleton is famous, and sold crystallised minerals from his shop in the museum. When he later moved to Stockport, to become curator of the town's museum, he took the window with him, where it still is. Pennington tried to sell the Castleton Museum, and eventually its contents were purchased, and are now displayed by Bolton Museum.

Windy Knoll Cave is near the spot where Boyd Dawkins and Pennington found thousands of extinct animal bones, now in various museums.

In the twentieth century, the Committee for the Archaeological Exploration of Derbyshire Caves did a lot of work in East Derbyshire, around Creswell Crags, under the supervision of Leslie Armstrong from the 1920s to the 1950s. Also, groups of local enthusiasts, notably the Peakland Archaeological Society, carried out further digs at caves in the Derbyshire uplands. One of their excavation directors, Don Bramwell, published a number of papers, and a popular book, *Archaeology in the Peak District*. Derbyshire cave surveys have also recently been carried out as academic research projects by Sheffield University's department of archaeology and prehistory.

The Blue John spar window made by Tym, and on display at Vernon Park Museum, Stockport.

5

GROTTO OF THE GODDESS

SACRED WATERS

The town of Buxton, at 1,000 feet above sea level, is dominated by hills, but by far the most imposing is Axe Edge, bulking impressively to the South West, with its peaty moors and whale-like back rising to a height of over 1,800 feet. It is from this bleak wilderness, now with the Leek road climbing its flank, and a perpetual convoy of stone and container lorries crawling like insects along the thin ribbon of the road, that the glacial melt-waters of ancient ice-sheets once poured their torrents down to the yielding swallets at its margin with Stanley Moor, where gritstone meets limestone, and the surface water dives down potholes, and rushes underground. In snowy winter conditions, Buxton is frequently cut off from the outside world, and Axe Edge assumes again the frigidity of a glacier. Even in summer there is an immense grandeur about this hill, and a sudden mist, and a chilly wind whipping across the rippling fronds of lumpy tussock grass can bring a sense of eerie isolation.

It is no surprise that this watershed of the prevailing, moisture-bearing, westerly winds gives the source of the waters which have made Buxton famous. One of only two thermal spas in Roman Britain, (the other is Bath) the springs' healing, aquatic properties were well known before the legions tramped into this inhospitable, but mineral-rich, part of sodden Britannia. The waters from Axe Edge and other hills which ring the town percolate down, down, into the very bowels of the earth, in a journey believed to take them close to the source of ancient volcanoes which once shattered parts of the Peak. Thrust up again by thermal pressure, the mineralised waters emerge at what was named by the native Britons, and recorded by the Romans, as *Aquae Arnemetiae* – literally, 'the waters of the goddess of the sacred grove'. No fewer than nine warm springs surfaced in the vicinity, giving a pure, almost taste-free spa water, unlike the sulphurous beverages so disliked by invalids at other spas.

Looming over the western fringes of Buxton in the snow, Axe Edge provided the melt-waters which carved Poole's Cavern.

The elegant Crescent created by the Duke of Devonshire for visitors taking the famous spa waters. St Ann's Well still gives of her bounty here.

Here, the Romans built a bathhouse, and a temple to the goddess, some foundations of which still exist under the eighteenth-century Crescent, near St Ann's Well – so named because a piece of Roman masonry was dug up nearby, inscribed AN.., probably nothing to do with Ann ! In fact, their bathhouse survived, and was visited by those seeking a cure from sickness throughout the middle ages, when it was regarded as a holy well, and until the eighteenth century, when the landowning Duke of Devonshire decided to build his spanking-new Crescent Hotel on top of it. He also chopped down the remnant of the sacred grove of ancient trees, which had survived since Celtic times. The well is still there, moved over the road from the site of the Roman bath, but still issuing waters freely to the citizens of Buxton, who queue to fill their containers – a right conferred by an ancient law of the town. The jet of water from a brass lion's head steams in cold air at a constant temperature of 28° C. The sacred grove is commemorated by an old coaching inn, the Grove Hotel, across the main road from the Crescent, which has the distinction of having an arcade of canopied shops at ground floor level, and the bar and dining rooms upstairs. In this vicinity once stood the old pilgrim chapel of St Ann. Daniel Defoe tried the waters during his early eighteenth-century visit, and commented not only on their pure taste, but the advantages of bathing in the lukewarm water. 'After the first coolness you find a kind of equality in the warmth of your blood and that of the water and can hardly be persuaded to come out of the bath when you are in.'

EARTH MAGIC

The other natural wonder for which Buxton is famous, and which provides us with evidence for the activities of ancient man in this place, is the enchanted underground temple of the goddess herself. Poole's Cavern, one of the original Seven Wonders of Derbyshire, does not have the grand entrance portal of Peak Cavern or Thor's Cave, but it has a magic of its own. Before the entrance was widened in the mid-nineteenth century by gunpowder, visitors had to crawl on hands and knees over glacial sediments, with the rocky roof pressing down on them. But once inside – what a spectacle! The first chamber was said to have been the refuge of an outlaw Poole, for whom the cave is named. Like the old entrance, it is still packed high with glacial mud and clay, which is festooned with archaeologists' numbered pegs, as countless finds have come to light. This was named the Roman Chamber by Prof. Boyd Dawkin when he excavated here in 1890, finding extensive evidence of the work of a Roman bronze smith who seems to have been churning out jewellery and trinkets for some sort of tourist trade – perhaps as offerings for the goddess herself, at the sacred grove, or in the cave itself?

Mid-Victorian visitors to Poole's Cavern were the first to benefit from the widened entrance, blasted by gunpowder.

Pegs still mark archaeological deposits in the glacial mud inside the 'Roman' chamber, and much work remains to be done when funding is available.

Further in is the River Sink, where the infant River Wye gushes between rocks, through a truly cavernous gullet of a cave, then disappears, to re-emerge in daylight at Wye Head spring, its official source, before it rolls down through the Pavilion Gardens and the centre of the town. A massive stalactite, officially the largest in Derbyshire caves, at two metres, hangs like a pendulous epiglottis from the roof. Known as the Flitch of Bacon, because it's said to resemble half a pig, it was vandalised, and the tip broken off in Victorian times, by rock-hurling visitors. The broken tip is now placed nearby, on flowstone-coated rocks below the footway, to illustrate how long it might have been before modification. Another feature is the Constant Drop, an eye-like flattened stalagmite on the floor of the cavern, where the rapid dripping has drilled a 3 cm hole. The view down this chamber from the bridge is spectacular, at 400 feet, the longest of any cavern in Britain, with cathedral-like arches, and glistening rock above and below. With what awe must early man have entered this chamber carrying flaming torches or flickering lamps, while the grotesque rocks cast strange shadows?

This cave was certainly known to early people. The first custodian when it opened as a show-cave in 1853, Mr Frank Redfern, began to make discoveries in the cave entrance as soon as he started excavations to enlarge it. Many of the objects he found were purloined by Thomas Bateman for his own private museum, but Mr Redfern opened a small

The main chamber of Poole's Cavern displays the Flitch of Bacon, with its vandalised tip still the largest stalactite in Derbyshire.

A constant drop has drilled an eye-like hole in this flattened stalagmite. Sticking your finger in and wishing is supposed to bring luck.

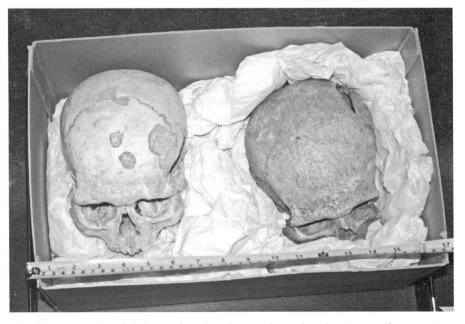

Skulls and remains of skeletons found in the cave by archaeologists are of uncertain date - anywhere from Neolithic to Romano-British.

Named the 'Poached Egg' chamber, this area has some very impressive stalagmites.

The cavern is arched with flowstone formations.

museum of his own near the entrance to the cavern. In it, he placed flint flakes, Roman brooches, human bones, and at least one skull. Even after excavations by Boyd Dawkins in 1890, Dr Don Bramwell found human bones from up to seven people, and items from Bronze Age and Neolithic times, when he conducted a dig for the Peakland Archaeological Society in the 1980s. Many of his finds, including Roman and Iron Age artefacts, are on view at the Cavern's visitor centre, or in Buxton Museum.

But the next chamber as the path mounts steadily upwards is bizarre. Should early people have required proof that inanimate rock could emulate living nature, and send them a symbol, here it is. The so-called poached egg chamber is festooned with stalactites, but it is the *stalagmites*, rearing up in every conceivable position on the adjacent rock that rivet the attention. With their reddish tips caused by minerals, or bacteria in the lime-rich water, they form columns of every size from small to colossal. I leave it to the reader to determine whether 'poached egg' is the best description of their appearance. Their growth may have been accelerated since the eighteenth century, when lime was dumped on the hillside above in vast quantities by lime burners, and is said to be at the rate of about 1 cm a year. The largest are thought to have taken just 400 years to reach their present size, and one is even beginning to form on the modern steel handrail of the walkway. If ancient humans – whose religious practices usually incorporated fertility rites – had penetrated this passage and seen shapes like these looming in the darkness, they certainly would have thought that the goddess whose warm springs issued in the valley below was trying to tell them something! Simulacra in nature, frequently the inspiration of cave artists to develop a theme from a suggestive shape in the rock, have been revered by primitive people the world over as messengers of the gods.

QUEENS AND GODDESS

Continuing deeper into the cavern, the passage walls close in to form a tunnel sixty feet high, with a flat bedding plane roof showing rippling from the violent flooding following the melting of glaciers. A great dome feature in the roof, nearer the entrance to the cavern, was also created by whirling floodwaters grinding rocks and sediment against the roof, and almost drilling through to within a few feet of the surface of the hill. The walls of this narrow passage are liberally coated with fantastic calcite flowstone formations. Calcite is limestone which has been dissolved by acidic rain/groundwater, which then deposits as crystals on the sides of the cavern as it flows downwards, releasing carbon dioxide gas. At this point in Poole's cavern, pure white calcite forms a simulacra of a 'Frozen Waterfall', and a little further on there is the 'Cat'. Folded ribs of calcite, said to resemble organ pipes, arch over the path, forming a false ceiling. At Mary Queen of Scots' pillar, Elizabethan cave visitors ended their journey, and left names and dates inscribed on a sheet of flowstone resembling an unfolded parchment. One is by the Scottish Queen's portrait painter, artist Rowland Lockey. It is said that Mary herself penetrated this far on one of her five recorded visits to take the waters at Buxton. She was staying at the Old Hall (then called the New Hall), situated next to the baths, in an effort to ease her rheumatism, caused by years of incarceration in draughty castles and manor houses on the orders of her cousin, Queen Elizabeth.

Since the opening up of the next chamber in Victorian times, we can now explore further than the ill-fated Queen, who is said to have scratched a poem on a pane of glass with a diamond ring, extolling Buxton's fame for salubrious waters and wondering whether she would ever return there. Alas, an unlucky fate was to lead her to Fotheringhay, and the headsman's axe. Around a bend, the footway halts at a boulder choke fifty feet high, with ribbed walls overhanging a most curious formation. One might almost be standing in an enormous womb, looking at a giant, partly-formed embryo, nestled beside the rushing waters of the infant Wye. Boulders have been coated by flowstone to create a giant cauliflower

Looking like melted cheese, some of these are named the organ pipes.

effect, christened by schoolchildren 'The Sculpture.' It is a suitable motif for the innermost sanctum of the goddess. Beyond here, boreholes have allowed remote cameras to explore a further series of chambers, never yet accessed by cavers, believed to continue for another one and a half miles, and full of interesting formations!

Remains discovered both inside and outside the cavern included prehistoric pottery, Neolithic hammers, and axes, some traded from as far away as Cumbria and Cornwall, and early Bronze Age artefacts. Stone axes are fairly common finds around the Buxton area. There are probably older remains at deeper levels in the cave yet to be uncovered. On Grin Low, the hill above the cavern, where Solomon's Temple, a late nineteenth-century tower with magnificent views now stands, was once a Bronze Age burial mound, where several crouched skeletal burials were found, including the impressive skull now the centrepiece of the cavern's exhibition area, which greets people with its 'Grinlow grin' as they walk into the exhibition. The name, which occurs at several places in Derbyshire, derives from *Grin* or *Grimr*, thought to refer to the pagan god Woden or Odin of the Anglo-Saxons and Norse, and *Low*, the name for a hillock or burial mound. The owner of the skull was probably a Bronze Age chieftain. We know this because the lower orders weren't interred in burial mounds, unless as

Thought to resemble a roll of parchment, this drier corner has many signatures carved by famous visitors.

Mary Queen of Scots visited Poole's when staying in Buxton, and got as far as the pillar named after her.

sacrificial slaves to the chief in his afterlife, their bodies burnt to ashes, and put in a pot at his feet, perhaps with the remains of his dog, or a horse.

Why a bronze-smith was based at the cave in Roman times is still a subject of speculation, but the shapes of some of the brooches he was making give us clues. These include dolphin, seahorse, chariot wheel, and trumpet – all suitable emblems of a goddess of the waters, who rode her chariot through the waves drawn by seahorses, and blowing her trumpet to summon her loyal school of dolphin guides. The Wye, into whose waters run the thermal springs of the Goddess of the Grove, is an enchanting stream, but those who wish to savour its mystery must trace it for some distance beyond the clamour of Buxton's water gardens, its inglorious culvert beneath the car parks of the shopping precinct, and its brief emergence through Spring Gardens, and follow it past Morrison's supermarket, the sewage treatment works, and the last outlying council estate. Here it enters Wye Dale, and the limestone cliffs rear up on either hand as, gathering speed, the stream rushes towards Chee Dale, where the pace slows, and deep in a secret valley, below the now-disused railway, it winds through a sylvan oasis, seen by a few and unsuspected by many. Here, where massive overhanging cliffs trail fronds of creepers over stepping-stones in lieu of a pathway, the chilled river calmly meanders its

The Cauliflower, or Sculpture, is at the end of the public tour, but further secrets lie beyond.

This tough-looking character greets you on entering the exhibition. He was a Bronze Age chieftain, buried at Grin Low, on the hill above Poole's Cavern.

way far from busy roads and madding crowds, beneath lofty Chee Tor, and some sense of the magic of the goddess can still be felt. One might envision her riding her chariot, with splashing escort of leaping dolphins, rejoicing in her waters on their way to the sea. The echoes of her silver trumpet fade among the high cliffs and silent woods, but the fauns, nymphs, and satyrs have heard her call, and are reasserting their right to this place.

Early people built a settlement on the hill above Chee Tor, and remains of their homes and field systems can be discerned in the bumps and hummocks. Across the vale at Wormhill is the legend of a dragon, which wrapped its coils around Knott Low, and here the last wolves in England were said to have survived longer than anywhere else. John Ruskin wrote longingly of the vale:

> There was a rocky valley between Buxton and Bakewell divine as the Vale of Tempe, where you might have seen the Gods and all the sweet muses of the light – walking in fair procession on the lawns of it... but you enterprised a railroad through the valley, the Gods are gone and now every fool in Buxton can be at Bakewell in half an hour and every fool in Bakewell at Buxton.

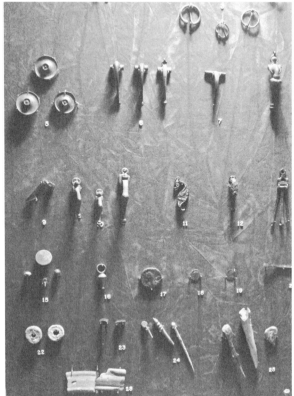

Above: A bronze-smith at work, in the entrance to Poole's Cavern in Romano-British times.

Left: Bronze objects from Poole's, and other caves and sites in Derbyshire, at Buxton Museum.

Happily, the railway is now a series of pathways, and no longer intrudes on this sylvan paradise of nature. Various plans for its reinstatement have so far come to nothing, as the series of bridges and tunnels, requiring expensive overhaul, make costs prohibitive.

The above fanciful depiction of the goddess shows what a slight deduction from common Roman jewellery motifs may achieve! Far more likely is that the identity of the goddess was one more local manifestation of that female tribal divinity of the Northern Britons known as *Brigantia*, and after whom the tribal confederation of the Brigantes took their name. It is said she had some influence as Britannia, based on Roman references in a bas relief showing the subjection of a bare-breasted woman. She is closely associated in character with the Irish Brigid, and is believed to have a special connection to rivers and wells, healing, poetry, wisdom, and the protection of her people. Curiously, one other blessing she bestowed was a patronage of smithcraft and metalworkers (hence the bronze smith at Poole's Cavern?). She is also a goddess of lofty places, and can be said to be the goddess of highlanders, in the sense of those dwelling in, and among, the hills.

The enchanting Wye glides between rocky overhangs in romantic Chee Dale.

LITERATURE AND LEGEND

To return to the name of Poole's Cavern, various theories have been put forward to account for it. Most agree that Poole flourished like a late medieval Robin Hood, living in the environs of the Peak Forest around Buxton in the fifteenth century, and according to the philosopher, Thomas Hobbes, who visited the cavern in the early seventeenth century, Poole was 'a famous thief, by spoils of those he robbed he used to live and towards his den poor travellers did deceive.' Most of the early fifteenth century was a time of turbulence, with the overthrow of Richard II by Henry Bolingbroke, the revolt of Owen Glendower against the usurper king, Henry IV, the wars with France, and the long years of the Wars of the Roses, when York and Lancaster vied for the crown, culminating in the bloody battles of Wakefield and Towton, fought within fifty miles of Buxton. All this rather put the Sheriffs out, and law and order suffered, particularly in the Royal Forests, which weren't really in their jurisdiction, except by Royal warrant – but who to apply to for one, as either contender for the crown could label you a traitor if you sent to the other!

It is believed that Poole, 'a man of great valour', being outlawed, maintained himself in this manner sometime between 1400 and 1460. There was a Pole family at Poole Hall, Hartington, not many miles away, who owned land around Buxton. Another Poole was said to come from Poole Hall, Staffordshire, and was a known outlaw. A third candidate, William Poole, came from the Wirral, and was outlawed in 1437 for kidnapping a wealthy heiress, and forcing her to marry him. One of these may have lived 'the merry free life 'neath the greenwood tree' hereabouts, as the Derbyshire hills were notorious for outlaws, and most of the caves in the neighbourhood of Buxton were periodically inhabited as their robber dens. A bit like foxes or badgers, no matter how many were killed or driven off, another generation was sure to replace them! There are many relics from Poole's Cavern of this period, from rough earthenware to animal bones and bits of ironware. If they hid their ill-gotten gains here they have yet to be found – but treasure hunters please note, tiny silver coins from the medieval or Roman periods, though valuable at the time when not much silver was about, are hardly worth the sweat of digging them up nowadays. Please leave it to the archaeologists, who are used to grovelling in mud on hands and knees for peanuts, or a smile and a nod from the landowner!

A visitor to Poole's Cavern in 1627, Dr Richard Andrews, after having squirmed into the cavern 'like a crab', was awestricken by the vast interior, and scraped a poem with his penknife upon the wall.

Rimstone pools created by calcite give the main chamber a special effect, and perhaps gave the cavern its name?

A melting rock, a flowing walk,
A marble flood, a dropping chalk,
A stoney sea, a raining flint,
A weeping tomb and I am in't.

He was not the only literary visitor shaken or stirred by the experience. Thomas Hobbes, the philosopher and author of *Leviathan*, came, as did the playwright and contemporary of Shakespeare, Ben Jonson, and, later that century, Charles Cotton, Derbyshire squire and co-author of the *Compleat Angler*. Dr Samuel Johnson, the compiler of the first English dictionary, complained it did nothing for his gout, while Daniel Defoe claimed, in his *Tour through Britain*, that the dripping water of Poole's Cavern reflected from the candlelight of the guides 'as ten thousand rainbows in miniature.'

When the Redferns took over the cave in the mid-nineteenth century they introduced, firstly, huge candelabra, bathing the cavern in sufficient light to elicit heartfelt oohs and ahhs from their visitors, and then, in 1859, a series of gas-lamps – a revolutionary idea in show-caves at that time. The legacy can be seen in the rusted gas-pipes and speckled soot patterns on the walls. Surprisingly, they were still glowing in 1965, when Frank

Above left: Roman silver coins of the first century AD at Buxton Museum.

Above right: Frank Redfern in old age – the cavern's first custodian, appointed by the Duke of Devonshire.

Redfern's granddaughter retired, and the cavern closed. When it re-opened in 1976, the current owners, Buxton and District Civic Association, had installed 100 electric lights.

A bright new visitor centre and shop have replaced the old museum, some remnants and curios of which can be seen near the cave entrance, but the atmosphere remains. Many visitors have picked up curious impressions whilst touring the cave, and even volunteers who work there have had eerie sensations, particularly when alteration works are being carried out inside to improve the pathways and access routes. One worker was wheeling in a barrow load of cement when, as he put it: 'I felt as if some unseen force was preventing me from going in. I couldn't move, then some of the volunteers came out and it seemed to relax. I didn't want to go inside again for quite some time.'

GROTTO OF THE GODDESS

Right: Candelabra, then gas-lamps, illuminated the cavern for the first time under the Redferns' stewardship.

Below: A Roman oil lamp made of earthenware once lit the cavern's interior.

6

CASTLES AND CAVERNS

DEVIL'S ARSE

If you have never seen Castleton, you should go there. The small town lies at the foot of spectacular Winnats Pass, and is dominated by a Norman castle protected on two sides by limestone ravines, one of which terminates at the largest natural cavern entrance in Britain. Peak Cavern is about fifty feet high and 100 feet wide. Its broad entrance portal stretches back for over 300 feet – wide enough to accommodate an entire village of rope-makers in former times. Though the cavern entrance is normally dry, Peakshole Water flows out of a resurgence spring in the rock just outside the portal.

This whole area at the head of the Hope Valley is riddled with natural caverns and mine workings, dating back to Roman times and beyond. The largest caverns in Britain include three local cave systems: Peak Cavern, with twelve miles of tunnels which link through to Speedwell Mine; Blue John Cavern, with the largest vadose canyon cave open to visitors; and the recently discovered Leviathan and Titan, the biggest yet found. Mam Tor, the 'Shivering Mountain', broods at the end of the Great Ridge within an hour's stroll of the village, occasionally sending shivers of rock and shale onto the heaps of natural debris already accumulated at its foot – including the former A625 road, which now diverts down the old route through Winnats, thanks to landslips. On Mam Tor's summit are remnants of a bronze/iron age hill-fort, with spectacular views of the Kinder Plateau and much of the central Peak District. The mines and caves dot the limestone margin at its foot, and continue in the ridge that sweeps round towards Castleton in the South East.

Peak Cavern appeared on a map of 1250 by Matthew Paris, and again on Gough's map of England in 1360, along with Peveril Castle. The cavern was once regarded as Derbyshire's chief 'Wonder.' Quite what it meant to very early man we cannot say, as no regular archaeological dig has ever

Above: Custodian's sentry box at the massive entrance to Peak Cavern.

Right: An 1822 view of the ropewalks and a hovel at the entrance to Peak Cavern.

Just beyond the entrance terraces, the walkway descends to stygian depths.

taken place there. This is partly because it is the property of the Duchy of Lancaster, and the terms of its lease forbid such interference with the considerable glacial terrace in the entranceway, or anywhere else in the cavern. There must be extensive remains of both animals and man, for such a vast portal would have attracted both to find shelter within. A university geophysical survey recently indicated the remains of former habitations on the terraces, and the current custodian has opened up the ruins of one or two of the former rope-makers' stone hovels, built into the side of the six terraced levels cut into the cave earth in former centuries. Animal bones of uncertain date do turn up from time to time, and the custodian's dog and cat, keenly interested in archaeology no doubt, have done a bit of unofficial excavating themselves!

In medieval times, the cavern had the reputation as a meeting place for tinkers and gypsies, and this was utilised by the poet and playwright Ben Jonson, when he wrote a play for the court of James I, called *The Gypsies Metamorphosed*, which describes the legend of Cock Lorel, 'a prince of rogues' who once dined with the Devil in the cavern, and how the Prince of Darkness ate a dinner of civic dignitaries, 'which he blew away with a fart, from which it was called the Devil's arse.' The time of gypsy brigands must have been after the decay of the medieval castle in the fifteenth century. It had been built in 1086 by William Peverel, who fought at Hastings, and was given the lucrative lead/silver mining region to administer in the Forest of the Peak. The archaeologist Thomas Bateman believed, from lead seals discovered in the castle yard, that Saxon Kings had held a castle here previously, but the walls and keep now standing are all Norman work – the square keep having been added by Henry II, after a subsequent Peverel rebelled, and the King confiscated their lands and castle in about 1173. Held briefly in the Barons' Revolt against bad King John in the early thirteenth century, it was used as the administration centre of the mining district, and offenders against the forest laws were imprisoned here. Like the Roman fort at Brough a little way down the valley, the castle's presence clearly indicated the importance of the site from the point of view of its mineral wealth, but also possibly at a strategic crossroads in the Peak. The town of Castleton, at the foot of the castle hill, has a protective ditch and embankment still visible in places, and the entrance to the castle was up a long traverse of a vertical hillside, and across a dizzying drawbridge nearly 100ft above the mouth of Peak Cavern. Security was obviously an issue!

As a schoolboy visiting Peak Cavern, our youthful party were assured by the guide that prisoners in Peveril Castle, on the crag above, were sometimes dropped down an *oubliet* from the castle – a pothole, which emptied into the cavern and pitch darkness. Such a narrow entrance from Cavedale does pierce the roof of the 60 feet high great cave, but whether there is another direct link from the castle, which is now run by

At the rear, the castle keep is protected by steep Cave Dale, and dominates Castleton.

English Heritage, has yet to be established. Certainly it may have been an enlightened medieval approach to offender rehabilitation; given a sporting chance that the crippled and maimed victim could crawl in the right direction through a shallow flooded passage to the entrance, they would be unlikely to go shooting their bow and arrows in the forest again.

By the early seventeenth century, when Thomas Hobbes, and his patron the Earl of Devonshire, visited the cavern, its entrance was filled with 'the shrill cry of the cord-winders and the busy hum of their numerous twisting wheels.' They looked on the busy ropewalks with their high gallows-like wooden structures and tiny straw-roofed hovels, loaded haycocks to feed the animals, byres, and stables. A channel at the bottom of the slope on the eastern side of the vestibule fills with floodwater after very heavy rain, and an alcove here is called Swine Hole, after the pigs kept nearby – doubtless their particularly pungent excrement was carried away by the floodwater, and washed down the gorge from time to time. Hobbes and others commented on the smell of the place as being the origin of its Devil's Arse legend. There are thought to have been as many as six cottages, and about thirty people lived here full time, rent-free. Pottery shards and animal bones have been recovered from an old midden dating to the early 1600s. It is believed that the Whitingham, Dekin, and Marrison families were the longest cavern dwellers, and a descendant, Bert Marrison, the last

Hovel in cavern entrance repaired by custodian Richard Taylor.

rope-maker died, aged ninety-eight, in 1983, although his family had lived outside the cavern, in the village, since about 1845. Bert is commemorated by a plaque inside the cavern. The present custodian, Richard Taylor, has opened up a couple of the remaining hovels to show the living conditions of the troglodytes, and displays a stone with a slit which he believes was filled with urine and used to disinfect money during the plague years of the seventeenth century, when people came to buy rope and twine. He lives in the custodian's cottage, just down the gorge, near Russet Well, the main resurgence for the underground waterways of the entire cavern system, and is the closest to a modern counterpart cave dweller – on some winter's days seeing no one in the cavern except his dog, cat, and members of his own family.

Wives and daughters of the rope-makers depended, even in the seventeenth century, on income from curious visitors, and no sooner had tourists arrived on the threshold than the people rushing out with lighted torches to greet them, reminding one visitor of Hades as described by classical poets. The visitor remarked on the roughness of his guide, 'but handsome enough and girl enough, she was,' which made up for her manners! Early tourists, on penetrating the cavern's underworld,

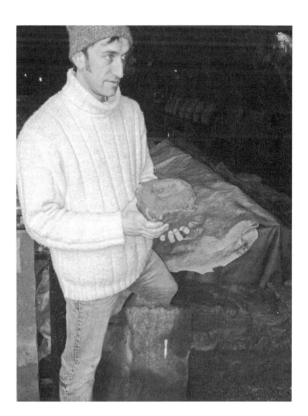

Richard Taylor and hollow stone he believes rope makers used to disinfect money with urine.

descended to make the crossing of the River Styx by boat, lying full-length on straw, while a guide pushed the boat, wading through water. The roof of the rocky tunnel descended to within inches of the tourist's nose, while candles flickered, and threw strange reflections from the walls and ceiling. As a young princess, Queen Victoria had been drawn into the cavern along this forty-foot tunnel, and thanks to her comments on emerging at the other end an alternative route was blasted from the rock, allowing subsequent visitors the luxury of 'Lumbago Walk.' This involves a stagger, bent double, for a similar distance, and the passage is still prone to flooding. The current custodian's sheepdog likes to pose at the end, just in front of the spotlight, as he emerges from swimming the Styx, and sends up a spray of iridescent raindrops of his own making. Just around the corner is the perpetual shower of Roger Rain's House, where a stream-way glitters down from Cavedale above, having found a fracture in the umbrella of volcanic toadstone, which keeps much surface water out of the cavern. Here is the bell-like aperture in the roof named Great Tom of Lincoln, and lofty voids loom all around, with twisted shapes and darkened corners. The legendary J. W. Puttrell of the Kyndwr Club first descended this chamber on 130 feet of Alpine rope from the entrance off Cavedale in 1902, to the amazement of assembled locals.

Princess Victoria was dragged lying full length in a boat through this water-filled, shallow passage.

High above your heads, pioneer caver J. W. Puttrell descended through a hole in the roof here, on a length of Alpine rope from Cave Dale.

Apart from one or two chambers in the roof, there is little in the way of flowstone, stalactites, stalagmites, or other twinkling calcerous deposits in the cavern, thanks to the lack of general percolating water from above. It is rather like walking through a giant plumbing system at this point, with every mark of the massive turbulence following the melting ice, when water crashed through these caverns with unbelievable force and terrifying volume. One can almost sense that, if the Devil turned some giant valve in the bowels of the Earth, it would all come crashing back again, sweeping us puny mortals away. Pluto's Dining Room is where the current tourist trail ends, at some railings close by the Devil's Cellar, where the tunnel plunges downwards into black, stygian depths over muddy rocks, and the start of adventures for the caving fraternity. Commodious and well lighted this chamber may be, but a less edifying spot for a hearty banquet would be hard to find. Somewhere in those miles of empty tunnels, Neil Moss, a young caver, lost his life in the 1950s, having been literally entombed alive. He is doubtless not alone.

The fear of a similar fate once induced a parsimonious baronet to part with more than he had intended, when one of the guides played the trick of 'accidentally' extinguishing the taper by which he lit the way in about 1770. Dekin, the guide, made off in the darkness with cries of terror, for

Members of Derbyshire Caving Club enjoying negotiating the miles of passages linking Speedwell with Peak Cavern.

he knew the waters were rising in the Styx passage, as an alarm pistol had just been fired. The baronet, who had been very mean with tips on a previous visit, now offered all manner of reward if Dekin would return and lead him to safety. 'This he did, making a merit of disregarding his own preservation for the sake of the baronet.' It served him as a laugh to the end of his days. By the next century, a scale of charges was posted in the village of Castleton, starting at one person with candles 3s 6d, two persons 5s – etc. Blasts of gunpowder in the largest chambers were 2s 6d each, and servants could accompany their masters and mistresses for 1s each. These charges were not cheap, and discouraged the poor and ignorant from pestering the cavern dwellers.

HOLLOW HILLS

Castleton, in addition to being uniquely situated and surrounded by interesting geology, is a very pretty village, with babbling brooks of clear water supporting a variety of aquatic life. Beside the main resurgence spring, Russet Well, a large, carved, Celtic stone head was found, now displayed in Castleton's visitor centre near the car park, and believed to be 3,000 years old. According to the legend, carved stone heads were linked to ancient practices of fertility, credited with averting evil and bringing good fortune. The ancient Celts hunted actual human heads, believing them to be the seat of the soul, and placed them near wells and springs as sacrifices to the water gods and spirits.

This is not the only fertility rite for which Castleton is famous. Every year on 29 May, the Garland Ceremony is performed in the village, when a 'bride and bridegroom' ride around the village on horseback, accompanied by a band, and watched by hundreds of townsfolk and visitors. They wear mid-seventeenth century clothes, and the bridegroom is encased in the Garland – a beehive-shaped framework covered in flowers, which hides him from view, apart from his legs and arms. They are preceded by Morris dancers, halting at each public house in turn, while drinks are sent out by the landlord. In the town's market place they dismount, and the Garland is hoisted up onto the small, square, medieval church tower to adorn one pinnacle, where it will remain for some time. Meanwhile, dancers prance merrily around a Maypole in the town square. A charity collection is made, and the collectors hand out fresh green sprigs of oak leaves to those who donate. In former times, they were said to nettle those who weren't wearing the oak leaf.

Stone head in 'Celtic' style from Russet Well, now in Castleton Visitor Centre.

A rare view of Castleton from above Peak Cavern, showing the steepness of the gorge.

The origins of the ritual are lost in the mists of time, and may commemorate the escape of Charles II by hiding in an oak tree from Cromwell's troopers. It is also claimed to have been introduced by Cornish miners who came to work in the local mines, but going further back, it is said to celebrate the rescue of a young maiden who was about to be sacrificed on Win Hill, a nearby pinnacle of rock, by men from Wormhill. The men of Castleton have marked their triumph by parading her through the town annually ever since. Maiden's garlands used to be carried at the funerals of unmarried women in Derbyshire, and some hang in the church at Ashford in the Water. The significance of placing the Garland over the pinnacle in Castleton is a clear indication of the subsequent fate of this lady, and coupled with the celebration of greenery in the oak leaves, is an obvious fertility rite. Given the proximity of Mam Tor, and its hill-fort held by the pagan Celtic Brigantes, and the likelihood that the 'men from Wormhill' were part of another tribal confederation, the Coritani, whose boundaries were contested hereabouts, these may be folk memories of a very distant event indeed. Two prominent opposing local peaks are Lose Hill and Win Hill, commemorating some all but forgotten conflict, in which clearly one army had the worst of it.

It was Sir Arthur Conan Doyle the, great Victorian storyteller, who said that Derbyshire was hollow. 'Were one to strike it with an enormous hammer, it would resound like a gigantic drum,' is how he vividly described the Peak. True to his description, the most enormous caverns

View from above Cave Dale looking towards Hope Valley and Lose Hill.

yet known in Britain are all clustered around the Hope Valley and the town of Castleton. We know that ancient people penetrated at least one of these caves at Treak Cliff, and left their remains. Probably they gained access to many more, and discoveries of their presence will be made as archaeological techniques advance. Later people, including the Brigantes and the Romans, set about mining the mineral veins, which outcrop in several caverns. By medieval times, Odin Mine, at the foot of Mam Tor, was regularly being exploited for mineral wealth, mainly lead, and by the eighteenth century, miners were extracting lead and Blue John, a unique decorative fluorspar, from caves in the cliffs above.

It seems that lead miners in the 1770s knew of a number of enormous caverns in the hills above the Winnats gorge, which probably contained abundant lead ore. A company from Stafford, Ralph Oakden and Partners, were persuaded to launch an ambitious mining project to imitate the Duke of Bridgewater's system for carrying coal by a canal network from his underground workings at Worsley, near Manchester. Their idea was that they could drive an underground canal from the foot of Winnats straight through to the first cavern, then on to the rest, linking them, and carrying the ore to the surface, and perhaps onwards to markets by a later canal, undercutting their rivals by bringing down the price. It was a good idea, and

Odin Mine is situated in a rift below Mam Tor, and is the oldest known mineral working in the area.

the Duke's former manager at Worsley directed operations at Speedwell Mine, as it became known. However, the rock was hard, progress, at a little over a yard a week, was painfully slow, and it was several years before the miners reached the first cavern, a stream cave eroded out of a mineral vein now known as the Bottomless Pit – so called because 40,000 tons of rock rubble was tipped into a lake at its foot without affecting the water level. Miners' old wooden steps, set into the rock, lead up from a platform at canal level to a mineral vein over sixty feet above, and the cavern recedes into darkness beyond that. Across the platform, the canal continues, but modern visitors don't get beyond here. It links eventually with a vast cave system, including Cliff Cavern, the recently re-opened Leviathan, and newly-discovered Titan – at 500 feet high, the largest natural cavern in Britain.

Mining ceased in 1790, when it was obvious that the £14,000 investment, which had only ever produced £3,000 worth of ore, was never going to pay! Tourists were already being taken down the 105 stone steps below the Winnats road, for a boat ride along the canal, by then. Propulsion was by pushing against the wall, or legging along the tunnel, but now it is by electric motor. A curious bit of graffiti in the mine pointed out to cavers who go beyond the Bottomless Pit Cavern is the 'Miner's Toast' carved on the wall, showing a bottle pouring into a

The entrance to Speedwell Mine is through the shop, once the mine manager's cottage.

Halfway House, on the underground canal, where passages meet.

wineglass, and the inscription 'A Health to all Miners and Mentainers of Mines Oct 20th 1781. J.I.B., M.N.' These are thought to have been Joseph Bennet and Matthew Nall, two miners who could write, and is in a dry alcove where they may have sat to eat their lunch! It has been established by divers that the 'bottomless lake' is only thirty feet deep, and full of rubble through which the water drains away. In flood conditions, it sometimes backs up to the platform above.

The Speedwell Cavern gift shop at the mine entrance is actually the former mine manager's house, and was previously a pub on the old coaching road up Winnats called The Speedwell. Among its collection of fossils, minerals, and mining relics now on display is an early eighteenth-century side saddle with a tale to tell. This belonged to a young couple, Alan and Clara, who in 1758 were eloping together, and heading for the Gretna Green of the Peak, the church at Peak Forest, where the vicar had a special licence to marry anyone who asked. Unfortunately, five rough lead miners at one of Castleton's inns overheard their plans, and saw that they were carrying a large amount of money. They followed the hopeful young couple up Winnats, robbed and battered them to death with pickaxes, hiding the saddle and bodies in an old disused mine in the pass. The murderers were never caught, and the evidence of a grisly crime lay undiscovered for ten

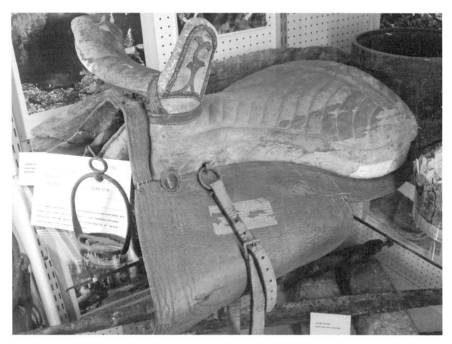

Clara's side-saddle, on which she eloped, only to be brutally murdered and her body hidden in a mine.

years. Eventually, the last of the murderers, after a long and painful illness, felt moved to confess his crime as he lay dying. He named the other four, who were already dead, but in a manner suggesting a sort of retribution had been meted out. One had committed suicide, another miner was killed by falling rocks, a third fell over a cliff, and the fourth went mad. Clara's side saddle is a reminder of this sad story -showing that plans don't always turn out as we would like them to.

A Cambridge university graduate, James Plumptre, recorded exploring Speedwell with a friend and two miners in 1793. The manuscript came to light in 1992, and solved the problem of what lay beyond the Boulder Piles, deep in Speedwell's farthest reaches. The miners had led Plumptre on a climb, past the boulders, into a vast, beehive-shaped cavern. This way is now blocked, but the description allowed cavers to excavate downwards from Over Engine Mine above, and after descending through some other blocked-off caverns with evidence of old mining activity, they reached the cavern now called Leviathan. It is in two parts, which amount to about 240 feet in depth. Over Engine Mine dates from the 1750s, and confirms that miners knew of the massive caverns in the hillside before the canal tunnel from Speedwell was begun.

A small passage full of miners' waste led off from the bottom of Leviathan which, on clearing, led into Far Peak Extension, normally only accessible by a dive along flooded passages from Peak Cavern's Far Sump. A further boulder-choked crawl of about 100 feet was excavated over several years, and in 1999 cavers were rewarded when they entered the lower part of Titan, nearly 500 feet high, and up to 300 feet wide near its roof. This was the biggest so far. There is no evidence of either miners or ancient man ever having entered here. Hourglass-shaped Upper Titan reaches to within 30 feet of the surface of the hillside at Hurdlow, on which one may stand unwittingly, with this vast space beneath one's feet. Conan Doyle was not wrong in his description of the hollow hills of Derbyshire. Many similar caverns are waiting to be discovered.

BLUE JOHN

The mineral now known as Blue John was claimed for a long time to have been discovered and mined by the Romans. Now it is said that the vases unearthed at Pompeii, once thought to have been made from the unique blue, purple and yellow banded fluorspar, were actually amethyst. A similar mineral to Blue John was obtained from Iran in Roman times, but Blue John is still thought to be unique. Most likely, if the Romans knew of it at all, it was because of preliminary mining work done by the local Britons.

An early caver plumbing the depths of Giants' Hole, near Castleton, one of the first big caverns found.

This seems feasible, if only because examples of local bituminous shale made into necklaces have been recovered from Bronze Age burial mounds in the Peak. This is an equally crumbly material, which needs reinforcing with resin and careful piercing for hangers to avoid shattering.

The main deposits of the semi-precious gemstone, Blue John, are in Treak Cliff, a ridge of reef limestone and its broken boulder beds, which lie immediately below and to the south of the ancient British bronze/iron age hill-fort of Mam Tor. Its two main sources from the eighteenth century were the Treak Cliff and Blue John cavern mines. There is an old shaft, now disused, which was claimed to have been the Roman entrance into Blue John Cavern; quite possibly, they searched here for lead veins, and found instead the decorative gem which can, with care, be moulded into vases, candlesticks, and jewellery. The Romans occupied other nations' territories, not to 'export democracy', but to exploit resources, just as the European colonial empires of the sixteenth to twentieth centuries did. They used slaves from the local population to provide labour and work mines. Folk memory claims that the village of Middleton, by Wirksworth, was founded as a Roman slave colony for workers in the lead mines nearby. The Roman fort of *Navio* at Brough, a couple of miles away, probably policed the mines around Castleton, and local Britons may also have suffered this fate. A similar principle was adopted by the Nazis in occupied territories to help pay for their war effort.

It is known that miners were working in 'The Waterhole', thought to be an early reference to Blue John Mine, by 1710. They were looking for lead, and much fluorspar was discarded, later to be re-excavated from spoil heaps when, in the mid-eighteenth century, Matthew Boulton started to make fashionable Blue John artefacts for the gentry. By the end of the century, there were thirty firms making Blue John ornaments in Castleton. It has sometimes been claimed that the Normans of Peveril castle named the stone, but now we are told that when the French started to import 'Derby Drop', as it was un-poetically known in the mid-1700s, they described it as '*Bleu et Jaune*', blue and yellow, after the stripes of colour in the crystal. The colloquial Castleton version became 'Blue John'.

Did ancient man know of these beautiful crystalline caves? It seems very likely, as at Treak Cliff in 1921, during opencast fluorspar mining, a small, sepulchral cave was broken open, having been sealed for millennia. This turned out to have been an original inlet for the tributary stream which once ran off Mam Tor, and helped form the caverns in Treak Cliff, before

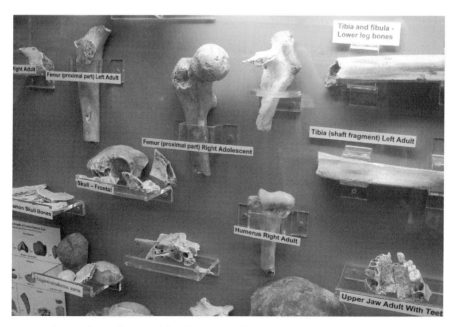

Human bones from the Sepulchral Cave are displayed at Treak Cliff Cavern café.

joining another stream which rose at the foot of Winnats. The first miner to crawl inside turned over a round object in the floor of the cave, and ran off in terror when he found he was holding a human skull! In this cave were the remains of three skeletons of late Neolithic date, together with a polished stone axe, an antler pick, a flint pebble, and animal bones. The presence of the 5,500-year-old axe and antler pick identify these people as miners themselves. Some bones can be seen on display at Treak Cliff Cavern, and others are at Stockport Museum. Beneath this cave, the entrance of which is now sealed, and partially obliterated by mining, miners discovered a new series of caverns, which they entered on a rope in 1926. The flowstone, stalagmites, and stalactites were so fantastic they gained the names for the newly discovered caves of Aladdin's Cave, Dream Cave, Wonderland, and the Dome of St Paul's. They are claimed to be the prettiest caves in Derbyshire. Treak Cliff is still mined in a controlled way for Blue John, and the objects made are sold in the cavern shop, and at Castleton. The view from the exit passage over the Vale of Hope, high in Treak Cliff, as you leave the caverns is one of the best in the Derbyshire Peak.

Blue John itself was created by oil from decayed organisms penetrating, and being trapped in fluorspar when it crystallised in the spaces between limestone joints and boulders. Geologists theorise that minute traces

A thick vein of Blue John spar exposed in Treak Cliff Cavern.

The huddled mine buildings at Treak Cliff, and the start of Winnats Pass beyond.

of uranium also trapped in solution distorted the formation of atomic structures in the crystals through radioactivity; this caused refraction, which affected the colours.

Blue John Cavern is open to the public, and is at the northern end of the limestone reef, looking across to the rugged outline of Mam Tor, and the great ridge which strides jaggedly up towards Lose Hill in the north east. It is possible to park on the remains of the old A625, which once ran past the entrance to the mine, and along a series of bends in front of the 'Shivering Mountain.' These now make an interesting walk, with large chunks of cracked and subsided road exposed, showing the successive attempts to reinstate by resurfacing, before landslips made it unfeasible and the route was abandoned to nature in the 1970s.

At the bottom of this hill route is the entrance to Odin Cave and Odin Mine, probably the oldest remnants of the region's industry, and worked in Roman times. Named after the Norse god of war and magic, Odin Mine was also referred to less romantically as Gank Mouth in old documents. The mine was mentioned in a record of 1280, when a poacher, John of Bellhag, was prosecuted for hunting 'at the entrance to Odin's Mouth.' At times, this mine employed up to 150 people, and was visited by tourists in the eighteenth and nineteenth centuries.

Stalactites and stalagmites in the Dream Cave at Treak Cliff.

Gank Mouth was the old name for Odin Mine. This is Odin Cave, visible from the road and 140 feet deep, not to be confused with Odin Mine at the top of the rift.

As late as 1827, the lead ore was said to contain a large proportion of silver. Convicts were employed to mine when slickenslides (explosive sections of rock under pressure) were encountered, which, when struck by a pick, could kill or maim the miner, sending flying splinters of rock bursting forth. The rift, which rises above the cave entrance, leads straight back, via a steep and dangerous footpath, to the shop and entrance to Blue John Mine. This rift once provided the main drainage off Mam Tor into the Hope Valley during successive ice stages of advancement and retreat, and as the overlying shales were worn down, exposing the limestone, water travelled underground, widening the joints, and creating caverns. Blue John Caverns are some of the largest of these, and in the 1780s the tourist entrance was down a miner's ladder of stemples, or beams, wedged into a sixty-foot-deep pothole. This dissuaded many visitors, and about 1836 a more convenient entrance was blasted into the hillside, with a series of ramps and inclined tunnels. There is an easy present-day approach to the cavern shop and entrance, along a causeway across the rift from the old A625, where there is a bus stop and car parking.

In 1765, the Barmaster recorded that Henry Watson, of Ashford in the Water, had sixteen stows or windlasses on Treak Cliff 'for anything that he could get'. What he wanted was Blue John, for his inlaid marble factory. But in about 1768, two other miners 'nicked' one of the disused mines on Treak Cliff, something they could do legally if a mine was not being

The hill pitted with mine workings in front of Mam Tor is Treak Cliff seen from a distance.

Neolithic skletons were discovered in a cave on Treak Cliff.

worked, and claimed the Blue John Mine as their own. It was in the hands of descendants for well over a century, and provided many of the best veins of the semi-precious stone. Some of the most colourful, distinguished by the banding, were found here, and many are worked out, though specimens can be found by picking through waste heaps – something the mine owners reserve their rights over. So don't try this, unless you enjoy fell-running from the enraged descendants of ancient Britons!

The present entrance is down a flight of steps with stone arching above, leading into the first cavern, with branch passages towards the vein workings, one called the Roman level, though no archaeological evidence for it being Roman exists. A series of switchbacks leads down the pothole. After the easy Ladies Walk is a narrow cleft, Fat Man's Misery, which opens into the Crystallised Cavern. Further on, down a staircase, is Lord Mulgrave's Dining Room, where an aristocrat once entertained the miners, who were showing him the wonders of the cave, to their meal breaks. The visitors' route ends in the Variegated Cavern – at nearly 100 feet high and 300 feet long, one of the largest show caverns in Britain open to the general public. The whole was still illuminated by candles until the 1960s, and was seen just as the early miners would have seen it! A sketch plan of the entire system was made in the early twentieth century, after exploration by J. W. Puttrell of the Kyndwyr Club. There does not seem to be a direct link to the other caves at Treak Cliff but the streams, which run through parts of the cavern percolate down through the boulder beds to emerge at Russet Well in Castleton.

7
ART AND MAGIC?

YOUNG CAVES

The magnesian, or dolomite, limestone, one of Britain's rarest rock types, outcrops the surface in a narrow band, never more than five miles wide, which runs along the borders of Derbyshire and Nottinghamshire, and into Yorkshire and Durham. It is younger than the hard limestones of the Peak District, being a mere 260 million years in age, and was formed in a shallow, tropical, inland sea as gently rolling dunes. These are now visible as the wavy strata of the bedding planes in the nine craggy gorges, which appear along the limestone belt between Doncaster in the north, and Sutton in Ashfield in the south. The most famous of these is Creswell Crags gorge, where a road has been diverted out of the valley to enable visitors to enjoy the unique archaeology, geology, flora, and fauna in peace.

Creswell's caves were formed beneath the water table (the level at which water gathers underground) in the rocky valley where they are situated, which was later deepened by streams from melting ice-sheets, finally making the caves accessible to animals and people. In geological terms the caves may be young, but the evidence of human activity there is ancient. The gorge began to be formed 500,000 years ago, yet the earliest evidence of people there is from only 55,000 years ago. This is not to say that even earlier peoples did not visit, merely that evidence from their time has not yet come to light. The difficulty of finding archaeological material at the deepest sedimentary levels is highlighted in excavation notes made by Leslie Armstrong, the archaeologist in charge of investigations there from 1922. Originally a surveyor, Armstrong became field archaeologist for the Committee for the Archaeological Exploration of Derbyshire Caves, and published extensive reports. One of his last reports concerns the excavation of Ash Tree Cave near Whitwell, a little to the north of Creswell, between 1949 and 1957. It is a good example of the conditions the archaeologists found.

Above: Peaceful Creswell Crags, home to waterfowl and cave hunters.

Left: Leslie Armstrong was criticised for not being a trained archaeologist, but he was professional enough to get paid for directing excavations.

...Ash Tree Cave is situated at Burntfield Grips near Whitwell. Excavations commenced there in 1949 and continued every season since. It consisted of a small chamber occupied by a mass of tabular fragments of rock loosely compacted by black loamy earth. The fragments appeared to have fallen from the roof, but at a depth of 1ft 6ins excavation revealed the stones had been piled there to cover deposits of human bones; typical Neolithic collective burials. The first of these comprised remains of at least two individuals, but were mainly those of a youth of slender build and good physique, aged 18 to 20 years. Much of the skeleton was present but no skull, mandible, or pelvic bones. To the right and 6ins lower was a second collection consisting of a clavicle, several phalanges and vertebrae a mandible of exceptionally robust type and the mandible of an infant; also a number of molar teeth. The mandible is abnormal in possessing only three instead of four incisor teeth and is from a young and strong man of about 24 years of age and represents an earlier burial than those first uncovered.

Beyond these collective burials and 16ft from the entrance the cave appeared to terminate in a fissure 1ft 6ins wide blocked at the base with stones and debris. Removal of the filling revealed a passage and 20ft from the cave entrance on the left side a cist of sub megalithic type was found. It contained the bones of at least two individuals deposited in a dry cove beneath the overhanging roof and enclosed by a semi-circular wall of limestone slabs, some placed vertically, to fill the space between the floor and the overhang and form a cist-like structure. The remains included the pelvis, sacrum, tibia a few ribs, numerous phalanges and vertebrae and a mandible retaining most of the teeth. The teeth are well preserved and exhibit considerable but not excessive wear. The mandible is a robust type believed to be that of a male aged about 30 years and is interesting in having a distorted left jaw apparently due to severe osteo-arthritis. The blocking up of this rear passage with stones seems to have been done deliberately after sealing the cist as additional protection. This burial clearly ante-dated the two previously found but not by any considerable length of time. In each case crumbs of charcoal and flakes of flint were present.

In the upper portion of the rocky debris and black loam which composed the floor of the cave prior to excavation, recent rubbish was found together with sherds of Roman and Iron Age pottery, "pot boilers" (stones heated in a fire, and added to pots to cook the contents) *and a few flint flakes. Artefacts of flint and bone, stone pounders and other evidence of occasional use of the cave were abundant down to the base of the Neolithic zone where the black loam merged into brown loam, which marks the Mesolithic horizon.*

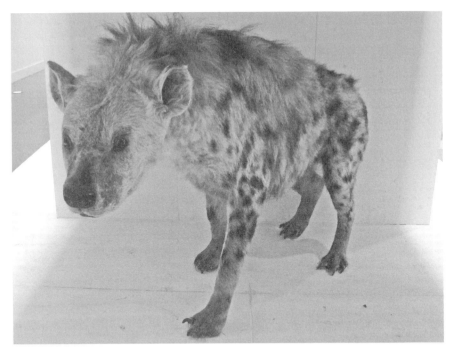

Hyenas like this one prowled the valley, and denned in the caves.

Sieving of the surface layers of the brown loam yielded to a depth of two or three inches evidence of casual use of the cave in early Mesolithic times. This consisted of a scatter of utilised fragments of split bone; stones suitable for use as pounders, sling stones; small split quartz pebbles; artefacts of flint and chert, a few microliths and microlithic flakes. The evidence was sparse and suggested temporary use. Patches of wood ash indicated campfires and there was a general scatter of charcoal crumbs over the area; probably spread by wind action. At the centre near the entrance a hole had been scooped to receive a deposit of incinerated human remains.

The brown loam layer was clearly of glacial origin. It was apparent that this loam had flowed into the cave in a sludged condition either as glacial drift or drift redistributed by the action of solifluxion (liquid mud). The loam in the cave is the same as that which masks the base of the cliffs bordering the valley and gives its U shape. The period of its introduction to the cave is proved by archaeological evidence as it rested upon and sealed a Creswellian occupation level and was capped by a Mesolithic one. It can therefore be correlated with the intensely cold and wet period which marks the Last Glacial 3, registered in the Pin Hole Cave, Creswell, by the deposit of the thick bed of stalagmite which sealed the Creswellian deposits there and marks the end of the Pleistocene (last ice age which ended 12,000 years ago).

The Creswellian horizon consists of the red cave earth of Pleistocene age upon which the brown loam rested and was a stratum 4ins to 8ins thick and throughout its depth contained evidence of casual occupation by man. Patches of wood ash indicated hearth sites and tiny fragments of charcoal were widespread. It yielded flint artefacts of the Creswellian type and the associated fauna included horse, bison, pig, reindeer, rhinoceros, bear, badger, hyena, fox, polecat, mouse, vole and lemming.

The underlying yellow cave earth and Mousterian (possible Neanderthal) horizon was 7ft 6ins in depth being compact, calcerous, very stony and exhibiting evidence of successive waterlogging; a creamy yellow in colour and inclined to be sticky when damp. The stones are tabular limestone derived from the roof and walls by disintegration, due to frost and water seepage. A central cross section of the deposit cut to the bedrock of the cave revealed three zones of Mousterian occupation comparable with those in Pin Hole Cave. Below 13ft from the surface the rock was reduced to yellow sand attributed to long submergence in still water and the consequent dissolving out of carbonate of lime.

In each zone the evidence of occupation was scanty and indicative of only occasional use either by man or animals. The artefacts found are similar in type to those from corresponding levels in Pin Hole but of inferior workmanship. Quartzite tools predominate, but few of them bear any secondary retouch and as a whole they suggest production for casual use. Flakes of flint or chert were present, but few in number. One well- worked flint point and a skilfully retouched scraper of black Derbyshire chert, are the most outstanding artefacts. Several bone awls were found, also portions of two bird tibias each perforated at the distal end for suspension and believed to be personal ornaments, or amulets. These are similar to examples found in Pin Hole in Mousterian levels. Utilised split bones, small pebbles believed to be sling stones and large quartzite pounders were numerous in each zone.

Upon the bedrock of the cave a hearth, approximately 2ft in diameter and 3ins thick was found. This was indicated by a concentration of wood ash and crumbled charcoal, enclosed in the calcerous clay, which forms the base layer of the cave. Splinters of animal bone were numerous at the base level and two heavy quartzite pounders were found on the floor, also a large portion of the humerus (leg bone) of a young rhinoceros. The fauna of the lower cave earth included all the animals recorded in the upper cave earth with the addition of mammoth, cave lion, cave bear, giant deer and red deer. Removal of a remnant of the hearth in the centre of the section revealed that it was over a fissure in the rock floor

18ins wide. This proved to be filled with burnt matter and humanly split fragments of bone in a matrix of sandy clay. The fissure was excavated to a depth of two feet beneath floor level but the bottom has not been reached. A Mousterian quartzite side scraper was found embedded in the fissure at a depth of 12ins and a deer antler awl, skilfully cut and trimmed at the point, was found beside the hearth on the floor of the cave.

Armstrong concludes in this report for the Derbyshire Archaeological & Natural History Society:

Though the material recovered in the excavation of Ash Tree Cave has not been large or spectacular it has enriched our knowledge of the Pleistocene people ... it duplicates and confirms findings at Pin Hole Cave and has provided new evidence relative to climatic events in the area at the end of the Pleistocene.

EXCAVATIONS AT CRESWELL

Archaeology was an infant science in the nineteenth century, and it was a keen amateur geologist, the Rev. John Magens Mello, vicar of New Brampton, who first became interested in Creswell after a quarry manager, Mr Tebbet, showed him some fossilised bones, which had come to light after livestock in one of the caves had trampled them out of the cave earth. The Rev. Mello sought permission from the landlord, the Duke of Portland, to begin an excavation at Creswell Crags, to which His Grace kindly assented. Rev. Mello was joined by the curator of Derby Museum, Thomas Heath, and in 1875 they attacked Pin Hole and Robin Hood caves with gangs of workmen, using dynamite to blast away the flowstone covering some of the cave earth! They also looked at Church Hole, where exciting discoveries have been made recently, and Mother Grundy's Parlour. Prof. William Boyd Dawkins, who joined them, started to systematically record the finds.

Creswell is now regarded as one of the most northerly sites visited by ancient humans during the ice ages. Early excavations resulted in the finds being scattered for exhibition among thirty museums, and investigation techniques were crude and destructive. Later archaeologists have re-examined the caves and spoil heaps excavated by the Victorian pioneers. A road has been diverted out of the valley, a derelict mill and cottages have been removed and a sewage works re-sited. Over £14m has been spent on a modern visitor centre housing many of the finds on loan from other museums telling the story of our ice age ancestors, interpreting the geology and landscape, and explaining the 'oldest art

ART AND MAGIC?

117

Above: Multi-million pound Creswell museum and visitor centre.

Right: Among the oldest bones on display are hippos from over 100,000 years ago.

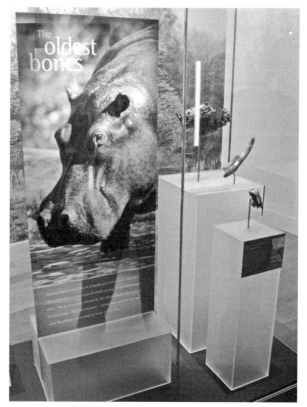

gallery in Britain.' The whole is run by the Creswell Heritage Trust, a registered charity, who conserve and interpret Creswell Crags and the surrounding landscape.

Visitors to the valley are today greeted by a tranquil scene of Crag Lake, home to swans and other waterfowl, and the crags ringing with the cries of game birds reared in the woods and adjacent flower meadows. Pin Hole Cave lies in the northern face, at the furthest end of the valley from the visitor centre. Its name comes from a tradition that Victorian ladies used to drop hatpins into a rock pool and wish. The cave extends 150 feet into the crags, beginning as a narrow fissure which widens out into a large chamber. The first known find from here was part of the palate and milk teeth of a young woolly mammoth, which drew the attention of Rev. Mello and Thomas Heath, who excavated 20 feet into the cave in 1875. Their excavation was methodically continued by Leslie Armstrong between 1924 and 1936 as far as the central chamber, working foot by foot, and recording the location of large bones and stone and bone artefacts by writing on the object. In 1936, Armstrong had progressed 80 feet into the cave, and had dug down fifteen feet. He left a twelve foot section of earth at the rear undisturbed for future investigation. Another

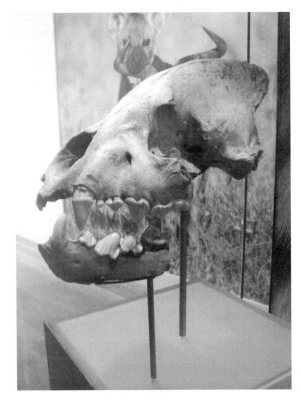

Massive teeth of hyena skull show its biting power.

excavation by later archaeologists in the 1980s uncovered a full hyena skeleton, and many small mammal remains. About twenty seasons' worth of work remains to be done.

Armstrong's discoveries revealed three phases of human activity during the ice ages: firstly by Neanderthals, which recent analysis places between 55,000 and 40,000 years ago, then by early Palaeolithic hunters between 40,000 and 28,000 years ago, and then by late Palaeolithic people re-colonising Britain after the intense cold about 12,500 years ago. The large bone deposits include remains of spotted hyenas which denned in the cave when humans were not in residence, woolly rhinoceros, wild horse, mammoth, giant deer and reindeer, many of which have been gnawed by carnivores. The large amount of small mammal bones from voles, shrews, and lemmings can be used to chart the changing climatic environment. But the most exciting discoveries at Pin Hole consist of the engraved pieces of ivory and bone found by Armstrong, reinforcing the authenticity of the engraved rib already found by Prof Boyd Dawkins in the 1870s at Robin Hood Cave.

On the end of a rib bone from a woolly rhinoceros was the engraved figure of what appeared to be a man with a semi-erect phallus. However, the phallus is now thought to be a natural mark on the bone, though microscopic analysis shows the figure is engraved. The figure is similar to art of the Magdalenian in south-west France. This and a broken ivory javelin-point engraved with wavy lines are thought to be 12,500 years old. A third piece, a fragment of rib, has been decorated with an attractive cross hatch pattern like plaiting, but dating this object is more problematic. The tools found include Neanderthal hand-axes of flint and quartzite, for chopping and butchering animals, over 30,000 years old, a distinctive tanged spear point, of the kind in use by early modern humans about 27,000-29,000 years ago, and late Palaeolithic backed flint blades, points, scrapers, piercers and burins, plus bone and ivory tools of about 11,000-13,000 years ago.

Mobile art found in Pin Hole – bone engraved with cross hatch plaiting.

HORSE'S HEAD

Robin Hood cave is the largest at Creswell Crags, and lies in the centre of the gorge. It extends a bit further into the cliff than Pin Hole, and has several chambers of varying size linked by narrower passages. It was first excavated by Rev. Mello and Thomas Heath in 1875 and by Mello and Dawkins in 1876. A row developed between Heath and Dawkins over the finding of the first example of portable art from a very early period in Britain, and the tooth of a scimitar-toothed cat. Heath claimed these had been introduced into the dig by Prof. Dawkins, but they are now believed to be authentic finds. An amber pebble, which found its way to Bolton Museum via Rooke Pennington's private collection, is also thought to be from Robin Hood Cave. Folklore persisted in attributing medicinal qualities and spiritual value to amber, and it was used as a wound salve until the nineteenth century. Hunters who visited Creswell may have picked up pieces of the fossilised tree resin washed onto the northern shores of Doggerland, between Britain and the rest of Europe.

A further dig in the 1880s by a northern antiquarian, Robert Laing, removed masses of cave earth from inner chambers, but little is known of his findings. In 1969, John Campbell carried out excavation of the entrance and scree slopes, and in the 1980s Rogan Jenkinson re-examined the cave to help analyse the Victorian work. At the lowest levels were remains of animals, including hippopotamus and narrow nosed rhinoceros, dating from 120,000 years ago, when the weather was similar to the present climate. Again, as at Pin Hole, it was discovered that Neanderthals had used the cave, and their distinctive chopping and cutting tools were found in the appropriate layers. Evidence for early modern humans, about 30,000 years ago, was found in leaf point tools used as knives or spear tips. Along with Pin Hole and Church Hole caves, there seemed to have been a fair amount of processing of arctic hare for pelts, as well as meat, about 12,500 years ago, and the cut marked bones of these animals are lying in sequence, with flint backed blades and bone points or rods. The bones of the hares were also used to make sharp awls for piercing leather.

The most exciting find from this era is the celebrated rib bone engraved with a horse's head, having a short, brush-like mane of the type of wild horse then roaming the grasslands. Just over seven cm long, the bone shows a high degree of polish, as if handled regularly. A series of vertical lines overlay the horse which have been interpreted as a possible stockade, and the back of the rib is criss-crossed with curved diagonal lines and squiggles. The piece was also stained with red ochre, and is similar in style to other examples from the period in Northern France and Belgium. Migrating hunters would regularly be travelling from these areas across

Doggerland, now covered by the North Sea. A hint of yet more art was discovered by cave art experts Sergio Ripoll and Francisco Munoz, led by archaeologists Paul Bahn and Paul Pettitt in 2003, when an enigmatic triangular shape on the cave wall was identified as a female symbol often represented in prehistoric art across north-west Europe.

PREHISTORIC ART

Previous to the discoveries on a sunlit April morning in 2003, it had been assumed that Britain was too northerly an outpost in the prehistoric world, and too close to the destructive effects of repeated glaciations, to hold any examples of cave art like those found in France and Spain.

This idea was overturned, when a team of cave art experts, using special high-powered lamps and side lighting techniques, found up to eighty examples of engraved and bas relief figures in Church Hole, one of the first caves they examined. The cave lies on the opposite side of the gorge from most of the other caves, and was first excavated by Mello and Heath, starting in 1875, after cattle stabled there had uncovered ancient

Robin Hood Cave, where the engraved horse head rib was found also housed Neanderthal tools.

fossil bones with their hooves. These amateur Victorian archaeologists broke through the flowstone layers covering the floor of the cave, and excavated six feet below this, uncovering masses of bone fragments. Heath commented: 'Out of a cartload (a full day's work) there were only four whole ones.' They concluded that carnivores had chewed them up.

Why hadn't the Victorians, including Prof. Boyd Dawkins, spotted the cave art, or indeed, anyone, since that time? One explanation is that since the floor of the cave was considerably lowered, the engravings were higher up, and difficult to see. Another is that no-one expected to see them, as the beautifully painted caves of France and Spain were then unrecognised, and in fact, the reliefs *are* difficult to see, even when pointed out today. Some, based around natural features in the rock, can almost be made out into any shape the viewer fancies. Rather like seeing figures in random patterns in curtains and linoleum. One is tempted to think that such art might be found in almost any cave in Derbyshire. I may have seen an auroch in Peak Cavern! However we accept the great experience of the experts. It is certain that some of the figures were created by man, but dating them is a problem. There don't appear to be any pigments in the Creswell rock engravings, and the main method of dating has been to estimate the age by analysing deposits of flowstone, which had partly covered the art, and its rate of uranium decay, which gave an age of about 13,000 years.

The cave art at Creswell is not in the same league as the elaborately coloured and moulded images found on the Continent. However, the Creswell figures are exciting, as they represent a native example of early creativity, the meaning of which is still not understood. It certainly was not just to make the caves look pretty for guests, though the superimposition of figures around the stag panel suggests rivalry among succeeding generations of artists. Archaeologist, David Lewis Williams, has argued that cave artists were Shaman magicians who attempted to literally draw some magic out of the cave walls by depicting the animals, usually in profile, which is the best angle to send a projectile into a beast of prey. There are few, if any, painted representations of people anywhere in Europe, and it has been theorised that the half human, half animal figures sometimes seen represent a clan totem, or a special affiliation with certain creatures, and that Shamans dressed in animal skins for magical dancing. Lewis Williams postulated the use of mind-altering plants and fungi as an explanation for the hieroglyphs and enigmatic marks made by narcotised, tripping-out artists, as a form of magical rite or initiation ceremony. But it may just have the same explanation as the graffiti of today – a simple egotistical statement of existence. I am here, look, I left this mark! Someone pointed out that the power of myth lay not in it making sense – myth rarely makes sense – but as it is repeated it gains strength in the repetition, and always finds believers.

ART AND MAGIC?

Above and right: Graffiti on this panel of rock at Peak Cavern left by visitors over the centuries is very similar to that on the next panel, taken at Pin Hole Cave, Creswell Crags. There may even be an ancient animal engraving in there somewhere!

Art doesn't have to have significance; like existence, it just is. Maybe that's what the cave artists had figured out.

Paul Bahn, who led the discoveries at Creswell, has been sceptical about the psychedelic origins of cave art, after originally being sympathetic to the idea. In a series of publications, he sought to prove there was little basis in Lewis Williams' ideas, and suggested that attempts at understanding the meaning of cave art may as well be abandoned. He even argued that narcotic drugs were not available to early man in Europe. There is a long history of controversy among cave archaeologists about prehistoric art, starting with the first discoveries in Spain in the 1870s, when an early 'expert' claimed they were all forgeries – which probably encouraged Heath to make his attack on Dawkins. Contentions of this kind have always been difficult to either prove or disprove!

On a positive note, the exciting possibility of finding further British sites has been suggested by Chris Stringer, director of the Ancient Human Occupation of Britain research group. In his book *Homo Britannicus*, he wrote:

> The Church Hole finds are significant for many reasons, beyond their age and what they depict. First, they suggest that careful searches of other British sites may reveal similar, or perhaps even more spectacular, examples of cave art. Second, they show commonality with European finds to the south and east, suggesting that human groups at this time were indeed linked across what is now the North Sea and the Channel. Third, this country has now been added to the distribution map of decorated Ice Age sites in Eurasia that stretches from Portugal to the Urals. The most northerly known decorated cave has moved some 280 miles from Gouy, near Rouen, to the British Midlands and the sparse portable objects at Creswell have at last been richly augmented.

CRESWELL ART

It is claimed that ninety figures have been identified at Creswell, mostly in the entrance chamber of Church Hole, where there is a lot of natural light – a few are in Robin Hood Cave and Mother Grundys Parlour. Up to fifty-eight of the figures are on the roof of Church Hole, and deeper in the cave there is a small panel of engravings in almost total darkness, where the artist would have had to use a burning torch, or lamp.

The engravings represent a wide variety of animals, including deer, horse, bison, bear, and at least two species of birds. Some areas show figures superimposed on top of one another. Much of the art would have

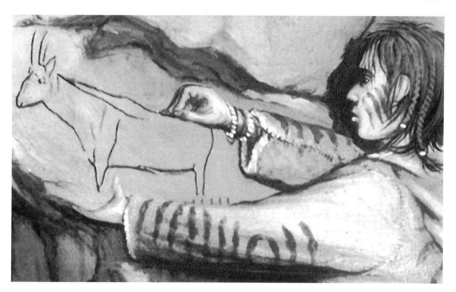

Prehistoric shaman artist, engraving with a flint burin at Creswell, recreated by artist Robert Nicholls. © www.paleocreations.com

been made using a sharp flint tool to engrave the outline of the animals in the rock. Bas reliefs were made by moulding existing features in the natural rock to create recognisable figures in three dimensions.

After the Victorian archaeologists tore through Church Hole in the 1870s, it was thought to be 'dug out', but it still contains unexcavated remnants against the cave wall, and in the entrance slope outside, buried beneath the spoil from the Victorian dig. A quartzite tool from Neanderthal times, 55,000 years ago, was recovered, and a selection of flint and bone artefacts from the end of the last ice age, 13,000-12,000 years ago. These include a serrated bone scraper for cleaning skins, bone awls and needles, and antler javelins. There were also remains of woolly mammoth, rhinoceros, horse, reindeer, and bison.

Church Hole is the richest decorated cave at Creswell, and the survival of this art probably relates to the relative hardness of the limestone, and its ability to withstand weathering, plus its lowland location, resulting in less precipitation percolating through the cave and potentially obscuring the engravings with flowstone. 'There is some debate concerning the total number of recognisable figures,' says the official guide.

> The degree of weathering that has occurred in the cave and the different rates of preservation resulted in some figures being more obvious than others... there are certainly around 20 images, perhaps more, but the ambiguous nature of the images will ensure that the debate will continue.

The stag is the largest figure, and was initially thought to be a goat because, comparatively recently, a visitor seems to have drawn a beard on its chin! A date of 1948 and the initials PM, engraved near the head, may be related to this modification. When the rock panel was carefully examined by experts, it was found to have the engraving of a red deer stag with antlers reaching up to the roof, and a natural solution hole used as an eye. At the bottom of the panel are three sets of vertical notches, with calcite deposits dating it to 13,000 years ago. The panel has many other superimposed doodles of animals. A little further into the cave is the bison figure, with its head worked up in bas relief from natural marks in the rock, and the ear and horns enhanced by the engraver. It has been praised for its magnificent proportion and fluidity by ice-age art experts. There are a variety of other shapes and lines on the ceiling and walls, including a long beaked bird, a horse head, parallel lines, triangles and scissor shapes which 'may be either flying birds or women.'

The problems of verifying and conclusively dating these markings, and determining whether they are natural or humanly modified is still a matter of opinion, and in the hands of the experts. I urge visitors to Creswell to retain an open mind, look, scrutinise, and come to your own fair and balanced conclusion. The thought that you are in the presence of art from an unimaginably distant past, created by people not unlike

Modern guide at Creswell demonstrates how to knap flint with a piece of antler in one of the caves.

ourselves, but so long dead that their bones have long since crumbled to atoms (unless we find a fossilised example) is a tremendously moving experience. There is absolutely no doubt that they left their tools here, and the evidence of them hunting and skinning animal prey is unquestionable.

There are a few other interesting caves at Creswell, notably Mother Grundy's Parlour, so named because a woman of that name lived there in the nineteenth century. It has a wide, semi-circular entrance chamber, and a deep, inner passage. Hippopotamus bones found here encouraged Mello and Dawkins to dig, again finding cartloads of animal remains, which date up to 130,000 years ago. Hyena coprolites (fossilised dung) were found in the cave deposits. Broken horse teeth show that people were extracting marrow from jaws here over 12,000 years ago. Radiocarbon dates from charred hazelnut shells and cattle bones prove the cave was used in Mesolithic times, between 6,000-8,000 years ago. Leslie Armstrong found flint tools near the entrance in the 1920s dating from the late glacial, 11,000-12,000 years ago, and also Mesolithic points and choppers. His quartzite Neanderthal tool finds have recently been questioned by British Museum expert Roger Jacobi, who felt that only one showed signs of manufacture. And so the experts continue to differ.

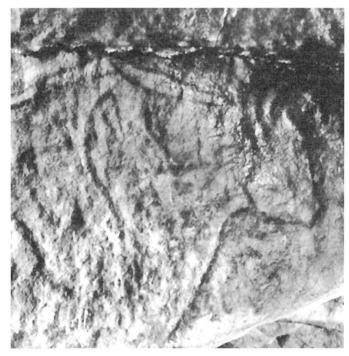

Undoubted prehistoric engraving of a bison, identified by cave art expert Sergio Ripoll in Church Hole Cave.

Above: Cave visitors at Creswell wearing hard hats and lamps prepare to pass the steel gates.

Left: The cave interiors are remarkably dry in spring and summer.

We can only guess at why this art was created, but walking through the gorge and visiting the caves, is like stepping into the shoes of Neanderthal and Late Palaeolithic hunters (supposing they had shoes). The caves still hold secrets, and those secrets are closely guarded, with grilles over the entrances to protect surviving deposits. Though we don't know the answers, perhaps future generations will solve some of the mysteries of our forgotten ancestors.

1. Stepping stones lead beneath limestone cliffs overhanging Chee Dale, beside the River Wye as it winds down the romantic vale from Buxton, where it issues from mysterious Poole's Cavern.

2. Overhanging rock-face provided shelter for prehistoric hunters and fishermen below Chee Tor. On the hillside above is an ancient settlement, which may date to Mesolithic times.

3. Jagged ridgeback of Chrome Hill is an ancient reef at the head of the Dove Valley, and is surrounded by swallow holes and small cave systems, with important archaeological caves nearby at Dowel Hall and Fox Hole.

4. Axe Edge above Buxton, at 1,800 feet, is a towering peaty moorland covered in tussock grass, and retains some of the bleakness suggestive of melting icecaps, even in the summertime. Its steep flanks are plied by trucks on the main Leek to Buxton road.

5. Beeston Tor, a high rock rises above the River Manifold, and is riddled with impressive fissures and caves, which are a magnet for serious climbers at all times of the year. Closer to the river lies St Bertram's Cave, where a saint hid from the world to contemplate God.

6. To reach this fissure high in Beeston Tor requires wading down the river, and fighting through undergrowth and bushes before a vertical rock-face, accessible only to climbers, is seen. Nimble, barefoot cavemen would have reached it with ease.

7. The church at Ilam, where the River Manifold returns to the surface after an underground journey of four miles, contains the tomb of St Bertram, a font with carvings of his life, and the remains of his stone Mercian teaching cross. Situated in the grounds of Ilam Hall, it was once in the village, but a squire had it moved to improve his view!

8. A fine winter's day view of Ravensdale Cottages in Cressbrook Dale, near ancient Ravencliffe Cave where Neanderthal tools were found. This remote, wooded valley runs into Millers Dale, just by the village of Cressbrook, where a former mill has been turned into trendy apartments.

9. Rocks above Ravencliffe Cave give some indication of just how high and inaccessible this spot is, even today, although in a south-facing alcove which benefits from the sun.

10. Fast food for cavemen? Ponies grazing above Ravencliffe look suspiciously at the photographer. Their ancestors were probably driven over these cliffs, and eaten by early man. There is little evidence of horses being domesticated before the Neolithic; instead, they were a major food source.

Above: 11. Cloud shadows chase over Wardlow Hay and Cressbrook Dale on a bright winter's day, with snow still showing on the high moors in the distance. The ancient fort of Camphill, now a gliding club, can be seen above the small village of Great Hucklow, on the wooded ridge to the top right.

Left: 12. Dove Holes Cave is the smaller of the two, and the most accessible, beside a bend in the River Dove near Nab Dale. Its oval form was probably created by swirling melt-waters when massive glaciers filled the valley, but it does not go in very far.

Above: 13. Dove Holes Caves are a magnet for passing walkers on the path beside the romantic River Dove. The larger cave has a steep rocky entrance, and again is quite shallow, with no deep passages leading off, although impressive to look at.

Right: 14. A spire of limestone at Ilam Rock is set at a funny angle beside the River Dove. Confusingly, it is not at Ilam, which is two miles away, just as nearby Dove Hole Caves are not at Dove Holes, which is even further! They are both in the lower reaches of Dove Dale.

Left: 15. Ilam Rock spire, viewed from Pickering Tor Cave across the valley. The cave is a small entrance behind Pickering Tor, which is another spire of rock facing Ilam Rock across the dale. Its archaeology is unknown, as like many caves it has not been dug.

Below: 16. Donkey Hole, or Radcliffe's stables, 30 feet above the River Manifold below Thor's Cave, once reputedly stabled a horse and hid a refugee from the Jacobite rebellion of 1745. It has a few partially blocked passages, and like most caves in the valley, was used by early man.

Right: 17. There are eight known archaeological caves in Cavedale, behind Peveril Castle, Castleton, but to reach them requires a steep climb up from the footpath in the deep valley below on slippery grass slopes – when using your hands, watch out for the stinging nettles at the top!

Below: 18. A clear view across Cavedale, towards the rear of Peveril Castle keep perched on its ridge with, beyond, the head of the Hope Valley, the start of Winnats Pass, Treak Cliff ridge, and Mam Tor hill-fort.

Left: 19. Stalagmites grow with mighty rapidity in Poole's Cavern, Buxton, on every available rock and crevice, because of the lime dumped on the hillside above, which percolates down, in a saturated solution of lime and acid rainwater which drips onto the end of each column, depositing tiny amounts.

Below: 20. The small stream was first dammed to create a watermill, and later a boating lake, which still gives the gorge at Creswell a peaceful feeling of tranquillity, punctuated only by the cries of waterfowl, or game birds from the surrounding woodland above the crags, which are riddled with caves used by prehistoric people.

Above: 21. The remains of the rope walks, which once also housed worker's small hovels, fill the entrance at Peak Cavern, also known as the Devil's Arse because, according to one early visitor, of the smell, as pigs, horses, and cows also lived here. The glacial terraces of mud and stones, although altered in recent centuries, have never been properly excavated, and may hold evidence of the ancient past.

Right: 22. Thyrsis's Cavern, the large West Window of Thor's Cave, allows light to penetrate the depths of this vast cavern, where Druidic rites were once held, near the village of Wetton, high in a rock above the Manifold Valley. A path leads out of here down the treacherous Thor's Cliff, but not for the faint-hearted.

23. A footbridge crosses the Manifold, below the spectacular rock which houses Thor's Cave on the hillside above. At many times of the year, the river here is entirely dry, having run underground through tortuous passages in the semi-porous limestone rock. But it flows on the surface when water is high in the winter.

24. The vast entrance portal of Thor's Cave is one of the most impressive in the Peak, and the worn rocks have been polished by generations of visitors scrambling over them. Excavations here showed that the situation was not lost on early man, and a buried skeleton was found.

Right: 25. The view from Thor's Cave along the Manifold Valley towards Wetton Hill and Ecton in late spring, when the trees are in foliage and the hawthorn blossom is out.

Below: 26. Which way to go? Ancient manuscripts tell that Demon's Dale was the entrance to Hell, but this information board at the start of the tiny dale simply says it was also known as Dimmins Dale, and doesn't go anywhere!

27. A walker pauses uncertainly, near strangely-shaped trees where Demon's Dale runs into Deep Dale. Perhaps he senses some unseen presence, or maybe the photographer asked him to pose for a moment!

28. The lower entrance into the area occupied by Horsborough Romano-British ancient settlement, which, like most old settlements in the Peak, dates back to the Bronze Age, and may have been a seasonal camp for Mesolithic hunters. It is just by the start of Demon's Dale.

29. The footpath leads off the road at the head of Winnats Pass, across fields to the archaeological Windy Knoll Cave, concealed in a hummocky old quarry below the hill-fort of Mam Tor.

30. Frost and snow caused a partial collapse of the roof near the entrance to Windy Knoll Cave, which is now closed to the public, as the notice proclaims. Near here, a fissure in a clayey waterhole trapped countless prehistoric animals, whose bones dated back to the Pleistocene.

31. No cave is too small, no crack too narrow, and no cliff too steep to prevent your intrepid author from investigating the remotest possibility of archaeological interest! Hole near One Ash Cave, Calesdale, off Lathkilldale.

32. *The Engraver*. Painting by artist Robert Nicholls. © www.paleocreations.com

8
GAZETTEER OF CAVES

The caves selected here are not a comprehensive list of archaeological sites in the region, but should offer features of interesting local history. Where a specific location is known, a grid reference is given. Some are not accessible to the public.

WESTERN FRINGES

STOCKPORT, though not in Derbyshire, is a meeting place of the Derbyshire Caving Club, and contains several cave systems of its own, in the soft, old,

A veteran caver explores water tunnels in sandstone beneath Woodbank Park, Stockport.

The town of Stockport sits on soft friable red sandstone easily eroded by the River Mersey. 1960s photograph near old Lancashire Bridge.

red sandstone rock underlying the town. The most natural of these were formed alongside the river valley, through erosion of faults in the rock by swollen rivers in glacial times. A stone axe hammer of Neolithic date was discovered near the River Goyt, in Woodbank Park, where a local cave well above the present level of the river is known by children as Devil's Cave, and has obviously been extended by the hand of man, and was probably connected with the eighteenth-century water tunnels. Another cave alongside the Goyt, further downstream, at Portwood, now hidden by undergrowth, was important enough to be marked on an estate plan of Portwood Hall in the seventeenth century, before the extensive water tunnels of the town were constructed in the following centuries. Another opens in rocks high above the river, halfway along Newbridge Lane. Others are beside Tin Brook, off Underbank, and at Brinksway, well above the River Mersey, are a number of hollowed-out caves, where natural erosion has been assisted by early Stopfordians' chipping away with stone tools at the soft, friable sandstone. Many have been in continuous use down to the present day. The most extensive systems are entirely manmade, and consist of a warren of water tunnels, excavated to bring river water from upstream to power early watermills in the town centre from 1750 to the early 1800s, rumoured to have been dug by French prisoners of war, some of whom are said to be buried down there. Part of one of these includes the Red Rock Fault, which widens into a large underground chamber close to the Market Place. Local legend has it linking church and castle across the Market Place in medieval times, and children claimed to have traversed part of it in the 1930s. In the late 1930s, five tunnel systems were also built to shelter thousands of residents from bombs during the air raids of the Second World War. One of these, on Chestergate, is now open for underground tours as a visitor attraction. There are others at Brinksway, Dodge Hill, Portwood, and Heaton Norris.

THE ETHEROW river valley to the north includes a known historic rock shelter, Pym's Parlour, or the Fairies Cave, in the sandstone beside the rocky gorge of the river at Broadbottom, which has yielded Roman remains and microliths, and seems to have had a religious use in Romano-British times. The Roman Fort at Melandra is less than a mile upstream. A footpath follows the rocky gorge from near the bridge in Broadbottom. At Tintwistle, where the first public reservoir system was built for Manchester with a series of dams in the late nineteenth century, there is a rift cavern in a gritstone quarry above the village, on the edge of moors which stretch across to Saddleworth Moor, where extensive remains of axe hammers and flint implements of late Palaeolithic or Mesolithic date have been recovered from beneath the peat.

KINDERLOW CAVERN GR070866 on Kinder Scout, above Hayfield, is a bit of a mystery. The entrance has been blocked and re-opened several times. Described as a series of fissures in the gritstone near Kinderlow End, with an entrance among tumbled rocks difficult to find, it may have been used as a seasonal shelter by hunting parties at this high altitude (1,800 feet), in warmer climatic conditions. There are the remains of a burial mound not far away. Some cavers claim to have found a way through to other caverns from here, but extensive rifts in gritstone are unusual, and the big cave systems are a couple of miles away, in the limestone to the south-east. A very energetic scramble is required to reach this height from Hayfield.

THE ROACHES are part of an extensive gritstone escarpment, stretching from near Hen Cloud and Ramshaw Rocks, above the main Leek, to Buxton road, with its weird shapes and fingers of rock pointing to the sky, including the notorious 'winking man', a rock that seems to blink as you drive by. The ridge stretches to Hanging Stone, near Wincle. At the end of the Roaches, near Hen Cloud, is a curious gritstone cave made into

The high and wild gritstone escarpment at The Roaches is visible for miles.

a castellated cottage called Rock Hall, built with boulders for the roof. Formerly, a gamekeeper used it to take money from day-trippers visiting the moors. Before him, it was home to Bess Bowyer for almost a century. The daughter of a noted moss trooper, 'Bowyer of the Rocks', once a terror of the neighbourhood, she herself sheltered smugglers, deserters, and thieves. A young woman claiming to be her daughter lived with her for a while, and was often heard singing among the crags in a foreign tongue. One night, she was carried off by strange men, and the old woman was found dead in the cave not long afterwards. The cottage is now used by climbers in this wild area, once famous for gypsies and pedlars. At nearby Flash, (the highest village in England), the term flash money was coined by the industrious forgers who resided there. Ludchurch Cavern GR987657 is at Back Forest, near Gradbach. Local legend tells of a family of cannibals, who used to eat travellers hereabouts in the eighteenth century! So perhaps it was unfortunate that the authorities suppressed the efforts of Walter de Ludauk to convert the local heathens to Lollardism, a form of enlightened Christianity, in the fourteenth century. Legend has it that he was preaching in secret, to a congregation gathered in the confines of Ludchurch Cavern, when some busybody Sheriff turned up with his men, one of whom, more eager than the rest, shot at Ludauk, missed, and tragically slew the preacher's beautiful young daughter, who died in his arms. The cavern is a high, open gritstone fissure which floods in bad weather, but has always attracted human activity, secluded high in the woods of a deep valley. It can be reached from the car park at Gradbach.

ALDERLEY EDGE MINES in Cheshire have been supplying copper ore since Bronze Age times, and it is believed that the blue dye in the skin of Lindow Man, found sacrificed in a peat bog at Wilmslow, was provided by copper salts from here. In the 1990s, ancient gold bars were found in the woods and declared to be treasure trove, and in the same decade 564 Roman coins of fourth-century date were discovered by members of Derbyshire Caving Club, in a blocked shaft at Engine Vein Mine GR 860775, and placed on display at Manchester Museum. Radiocarbon dating of timbers from the bottom of the same shaft showed it to date back to the century before the Roman occupation of Britain, and a wooden shovel discovered in one of the mines in 1878 was also revealed by the same method to be at least 3,700 years old, placing it firmly in the Bronze Age. The shovel had been discarded, then utilised as a prop by an amateur dramatic group in Alderley Edge, where it was rediscovered by Alan Garner, the playwright, who recognised its importance, and had it analysed. A landscape project has since identified Bronze Age hearths for smelting ore on the Edge, and microliths from a Mesolithic seasonal camp have been found.

The copper mine tunnels at Alderley Edge stretch for miles, including quite large caverns. Some shafts date to the Bronze Age.

Stone hammers and antler picks have also been recovered from the mines. These were first recognised by Dr J. D. Sainter, who found the shovel at Brynlow in the 1870s. Over a hundred stone tools, made from glacial pebbles, and used for pounding when lashed into a wooden handle, were discovered by Prof. Boyd Dawkins, who was the first to recognise ancient opencast mining on the Edge.

Currently, three mines, Wood Mine, West Mine, and Engine Vein, can be visited by the public, wearing miners' safety helmets and lamps, and guided by experienced members of Derbyshire Caving Club. The Club are engaged in an ongoing archaeological excavation of the mines, supervised by the County Archaeologist, and landowners including the National Trust. Lead, copper, and cobalt ores were mined here up until the early twentieth century, and the underground galleries are extensive, and reach all the way down to the village. A number of tragedies have occurred when unsupervised youngsters became trapped or lost in the tunnels, including flooded sections, and entrances have been gated to avoid this. There are several curious legends about Alderley Edge, which reflect its associations with an ancient culture of metal smelting.

Probably the most famous legend concerns the wizard. A farmer from Mobberley was taking his white horse to market at Macclesfield (in some

Wizard restaurant on The Edge is close to where the wizard met the farmer in the story.

versions it is black), confident of getting a good price. Near the Thieves Hole (behind the present Wizard restaurant), he meets an old bearded man in long grey robes who offers to buy the horse. The farmer declines, and proceeds to market, but fails to sell the beast, and coming home again, meets the mysterious old man near the Thieves Hole. This time he agrees to sell, and is beckoned to follow him into the woods, where the old man strikes a rock with his staff, which opens, revealing iron gates and a cavern beyond. The trio enter, and the farmer is amazed to see sleeping warriors with sword and helm, beside their sleeping white (or black) steeds, except for one, who has no mount. The old man, who of course is a wizard, indicates a pile of money and precious things, from which the farmer grabs as much as he can. He is told that these warriors sleep until summoned to a great battle in which they will decide the fate their country. Finally he is asked if he will choose a sword or a horn, and chooses the horn, which apparently is the wrong choice, as the warriors begin to wake up, the horses snort and stamp, a whirlwind fills the cavern, and he finds himself whisked outside as the iron gates clang shut, and the rock closes with a thud! The poor horse, of course, is still inside. The farmer never finds the entrance again, search as he will. This folktale is very similar to one told of a shepherd in Snowdonia, North Wales, in which case it's a bell, and not a horn, and another, told of the Eildon Hills in the Scottish Lowlands by Sir Walter Scott. Each place, however, has early associations with mining and metalworking. A white horse is the steed, in Celtic legend, of the horse goddess Rhiannon.

Stormy Point, on the rocky edge of the escarpment, nowadays actually has iron gates and bars over entrances to the mines, and below the Edge is the Wizard's well, with a face carved in the rock by Alan Garner's Great, Great, Grandfather, a stone mason, who it is said, also made the 'Druid's Circle' of stones in the woods. A poem, "Drink of this and take thy fill, for the water falls by the wizard's will", was carved onto the rock at a later date. Alan Garner has woven many of the local legends and places into his stories. There are several Bronze Age burial mounds in the area, and the remains of a thirteenth-century medieval castle built, or started to be built, by Earl Ranulph of Chester, to guard the rich pickings from the mineral veins, (as at Castleton), and for the position's strategic geography, not for the fine views, as has been suggested. In addition to copper ore, silver has been found locally. Discoveries of flints near Castle Rock have been linked to the idea that this cliff may have been used by prehistoric people to drive herds of game over the Edge, and butcher the slaughtered animals at the foot. There is evidence of a nearby camp of Mesolithic hunter gatherers dating back 8,000 years, but some of the flints may be from when the Castle Rock was used as a lookout by Civil War soldiers, who carried spare

Castle Rock, Alderley Edge, where prehistoric man may have driven animals to their doom, is close to an ancient settlement.

flints for their flintlock muskets. High in the woods are the remains of an Armada Beacon, and this consisted of a conical tower, until demolished in the early twentieth century because it was unsafe. As a beacon, it may have been much older than 1588, and was in use until the mid-nineteenth century as a precaution against invasion by the French.

DOVE VALLEY

DOWEL HALL CAVE GR075676 lies in a dry tributary valley of the Upper Dove. It is an old resurgence cave – that is, a cave which once pumped out water when the water table was higher due to nearby melting glaciers. Not far away, the underground stream which formed this cave system still resurfaces lower down the valley. It was in an attempt to reach this stream from Dowel Cave that members of Derbyshire Caving Club were digging in 1958 when they came across large numbers of bones, and called in Peakland Archaeological Society to help with the excavation. The immediate area is one of the most scenically remarkable in the Peak District, with Dowel Dale lying in the shadow of the limestone reefs of Chrome Hill and Parkhouse Hill, rising to resemble miniature Alps in

Dowel Hall Cave is approached from the road, up a steep bank.

their rugged, sharp ridges. The area has a number of genuine potholes called swallets, including the large Owl Hole at the head of Dowel Dale, a remnant of an ancient glacier's melt-waters. Excavations showed that no fewer than ten individuals, including six adults and four children, and the headless skeleton of a dog had been buried in three different sealed-off parts of the cave. Most of the bones were thought to be Neolithic, but radiocarbon testing of an antler point indicated it was 11,200 years old, which, if it had been in use as a tool, shows human presence at the start of the Holocene period, at the end of the last ice age. The skeletons were in good condition, and became distributed around various museums, including Buxton and Stockport, where two of them form a cave burial display. However, given the dating of the antler, it would be interesting to test the age of the human bones! Later artefacts found in the cave included flint flakes, a beaker shard, and Romano-British relics. The cave lies up a short, steep slope below the rocks not far up the dale from Dowell Hall Farm. Across the road from the farm, and seventy-five feet above the field adjacent to the lower resurgence, is Etches Cave, which is gated and believed to still contain unexcavated remains.

FOX HOLE CAVE **GR100662** on High Wheeldon hill, just outside Earl Sterndale, is in a remarkable position, near the apex of a seemingly conical

Alan Burgess beside Neolithic bones he discovered and helped excavate at Dowel Cave in the 1950s. They are part of the Origins Gallery display at Stockport Story Museum.

peak, yet so situated that it is nearly hidden from view until stumbled upon in a concealing dip. It is still gated because further archaeological remains await excavation, and it has already yielded some of the oldest evidence in the Peak. High Wheeldon is a continuation of the limestone ridges of which Chrome Hill and Parkhouse Hill form a part, not far from Dowel Dale. Visitors to the cave must climb the steep sides of the hill, which is on National Trust land, until nearly at the summit, then divert along the north-eastern flank until the small hollow which conceals Fox Hole Cave is found. It was here, in the 1920s, that a dog became trapped, digging in what was thought to be a rabbit burrow. The animal was freed by the destruction of the entrance capstone to what turned out to be a cave burial. Inside, the cave excavators found the bones of several individuals within an entrance passage, and two chambers. The later dig by Don Bramwell and Peakland Archaeology Society between the 1960s and 1980s uncovered the floor to a depth of over six feet, finding a polished Neolithic stone axe, human jawbone, humerus, and tibia. The two latter items were carbon dated to 6,000 years old, some of the oldest *human remains* yet found in Derbyshire. Andrew Chamberlain, of the Department of Archaeology at Sheffield University, believed they were part of an Early Neolithic tradition starting to utilise caves for funerary purposes, a practice less evident in the preceding Mesolithic period. Even more interestingly, two items of worked antler recovered gave late glacial dates of 11,970-12,000 years old, showing humans were present in the cave while the ice still held the hills in its iron grip. Other animal remains, including a bear's tooth, were left in situ. Later relics included Bronze Age pottery, Romano-British artefacts, and animal bones. The cave is in an ideal situation for hunters to scan the surrounding valleys for passing game from a virtually unseen vantage point, and then plan their routes for cutting off and trapping the animals in a killing zone.

FRANK I' TH' ROCKS CAVE GR131584 is near the junction of Beresford and Wolfscote dales, about a mile south of the picturesque village of Hartington, with its duck pond and old church. A muddy path leads across fields towards the start of the limestone dales beside a footbridge over the River Dove. The cave is hidden up the hillside beyond the most obvious cave entrance. Here excavators in the 1920s found the bones of ten people including two adults and eight children from the Neolithic period. There were also Iron Age and Romano-British artefacts. The archaeological remains are in Buxton Museum. Nearby Wolfscote Grange Caves are about fifty feet above the river, in a highly visible position below a cliff, and the raised lip of the main entrance is polished like marble by generations of people scrambling inside.

Road from Earl Sterndale heads into the dale below Fox Hole Cave in the summit of a conical hill.

Frank in th' Rocks Cave is in a south-facing alcove above Wolfscote Dale.

Frank in th' Rocks Cave is to the right, Wolfscote Grange Caves to the left.

REYNARD'S CAVE GR145524 is reached by a long walk further south along the Dove from Hartington, or a shorter walk from Milldale. The explorer will pass many caves, all of which doubtless have ancient history, but have yet to be excavated. After passing the spectacular Dove Holes rock shelters, which face other caves across the valley, near a bend in the river, the caves of Pickering Tor and Ilam Rock are reached, where another footbridge spans the river to allow visitors to admire these fine examples of Karst limestone scenery. After following the river along duck boards, the Lion Rock is passed, and looking up through the trees to your left, you will see a well-worn scree slope of broken rock, and, high above, a spectacular natural arch of limestone rock. This is quite a scramble for the unfit, and becomes steeper towards the top as you pass beneath the arch, where it is necessary to go hand over hand, clinging to well-worn boulders, and sliding on the clayey ground. Shortly, you will reach Reynard's Cave to the left, which has a flat patio area in front, ideal for prehistoric barbecues, and quite hidden behind the arch from the Dale below. It has a commodious and level interior where a hundred people might sleep uncomfortably side by side, but before you decide to take it,

Right: Magnificent limestone arch frames the approach to Reynards Cave.

Above: Reynards Cave interior is flat and commodious with a level patio, hidden from the valley by limestone arch.

scramble up a bit higher to the inconvenient Reynard's Kitchen, at the top of the clayey boulder-strewn slope you have just traversed. This cave has walls liberally coated in green slime, and a climb in the roof of interest to cavers. If you manage to get inside, the author takes no responsibility for readers who break their necks, or a limb, on the way out! However, it does have the advantage of discouraging school parties. Bronze Age and Romano-British remains have been found, and some are on display at Buxton Museum. There is also eighteenth-century graffiti. It would be strange if these two caves were not used by much earlier peoples, and there are some detached pieces of red cave earth waiting to be sifted. Returning to the valley bottom, Tissington Spires' impressive columns of limestone are just a bit further on, and if you make it over Lover's Leap, the Isaak Walton Hotel beckons at the end of Dovedale.

CARSINGTON AREA

Across the limestone plateau a few miles to the South East of Dovedale lies the area believed to have been the *Lutudarium* of the Romans – their major lead/silver mining district of the Peak, where slaves were forced to labour underground. Carsington Water, a large reservoir, now occupies the valley, but the villages of Carsington and Hopton lie just off the new road, which skirts the lake. The district is still riddled with old lead/zinc workings, and some caves.

CARSINGTON PASTURE CAVE GR243541 produced one of the most interesting discoveries in recent years, with a grotto of the dead stumbled upon by members of Pegasus Caving Club from Nottingham. The cave lies in a concealed hollow, high on a promontory of limestone between the villages of Carsington and Brassington, has a chimney shaft, and was used as a shelter by miners until the nineteenth century. After excavation, no fewer than twenty Neolithic skeletons (up to 4,000 years BC) were identified, including nine adults and eleven children, with radiocarbon dates ranging back 5,000-6,000 years on animal bone, including an extinct auroch. There were flint flakes, a bone pin, worked antler, cut-marked human bones suggesting defleshing, as at Gough's Cave, prehistoric pottery, and Roman artefacts. The caving club were digging down through the floor of the entrance chamber in 1998 when they broke into a second chamber heavily encrusted with stalactites and stalagmites, some of which had coated the human remains, which gave the name 'Yorick Chamber' to the cave!

Brassington was on a main stagecoach route – an old lead mining centre now visited by hikers off the beaten track.

Caver Andrew Walchester literally fell into the cave, his helmet lamp illuminating the horned skull of a goat, and a human skull sporting a stalagmite. University of Sheffield archaeologists photographed the interior prior to removal of the remains, apart from one skull heavily attached to the cave wall by a stalagmite. A pit blocked with rubble and more remains led down to a third, lower chamber. There is evidence of further chambers leading from the entrance, and more human skeletal parts have been recovered, causing experts to suspect more discoveries yet to come. The chambers seem to have been deliberately sealed with clay and boulders, as in many cave burials, and used for burials over a period up until Iron Age times. A *Time Team* dig and programme in 2003 about the cave and a nearby Bronze Age burial mound focussed national interest on the site. A public footpath between Carsington and Brassington passes close by.

RAINS CAVE **GR226553** at Longcliffe near Brassington, a sloping chamber with crawls leading off, was excavated in 1889 by J. Ward, and revealed the remains of six individuals from Neolithic and Bronze Age times, along with prehistoric pottery, flint artefacts, kite shaped arrowheads, and animal bones. The entrance is concealed by large boulders to the East of Longcliffe Crags, and is very difficult to find.

Impressive rocks at Harborough have attracted man for millennia.

HARBOROUGH CAVE GR242552 beside the High Peak Trail at Harborough Rocks is spoilt by the nearby mineral works with its belching industrial processes, but still well worth a visit for its impressive dolomite rocks. Here Daniel Defoe encountered the poor lead miner's wife and her five children described earlier. It has evidence dating to the Old Stone Age reindeer hunters, and smashed reindeer bones shows marrow was extracted in the time-honoured fashion to be consumed here. In fact, the successive layers of the cave floor have revealed remains of domestic dogs, deer, and wild boar, and at the lowest level, hyena. They have also yielded jewellery of the Bronze Age, gold rings, and a rare brooch set with coral. Remember, it is illegal to dig without the permission of the landowner, and all finds should be reported. The remains of about four individuals of Neolithic date have been recovered, along with leaf shaped arrowheads and later artefacts.

MATLOCK CAVES lie in an area bounded by Matlock, Cromford, Bonsall, and Snitterton, and are a prolific series of caves which have been worked by lead miners going back to the Bronze Age. A couple in Matlock Bath are still open to the public, and have histories going back to Roman lead mining and beyond. Jug Holes Cave GR279595, in a wood on Masson Hill, was singled out by English Heritage as having special archaeological interest. Apart from its colony of bats, it is renowned for the stickiest mud in Derbyshire, made up of decomposed toadstone, a volcanic rock. Extensively de-roofed by miners, the large cave entrance is in a private wood criss-crossed by footpaths near Leawood Farm. The cavern has passages half a mile long, leading downwards to a natural cave with a huge, flat, slanting roof, and big fallen rocks sloping upwards, covered in a silent, petrified cascade of snow-white stalagmite flow, sometimes dirtied by sediment and the boots of cavers. Apart from 'the sound of a barking dog' caused by a now-extinct siphon pool, I cannot discover what other natural or archaeological interest there is. A visit to the Derbyshire Mining Museum, in the Pump Rooms at Matlock Bath, is a must in this area, once a very fashionable watering hole for the eighteenth-century gentry, where the River Derwent cuts through a breathtaking limestone gorge.

MANIFOLD VALLEY

The Manifold Valley is just over the border in Staffordshire, and runs more or less parallel with the Dove Valley in a southerly direction, cutting through the limestone never more than a mile or two away from its sister stream, and with its tributary, the Hamps valley, contains the largest concentration of archaeological caves in the White Peak. The Hamps and Manifold are unique in the Peak, because their boulder-strewn beds are rendered completely dry for a great distance in spring, summer, and

With little rain, the River Manifold often runs dry beneath Thor's Cave.

autumn. At this time, the rivers disappear underground into tortuous chambers, which have never been entirely penetrated by divers. The Manifold rushes underground through a series of sink holes beginning near Wetton Mill, re-emerging at Ilam Risings resurgence near Ilam Hall, four miles away! The exposed bedrock in the river below Beeston Tor shows fossilised crinoids, once deposited on a tropical seabed millions of years ago. It is said that a sink hole near Wetton Mill was descended to 90ft by cavers in the 1920s who discovered an underground lake. Since then, the hole has been blocked by boulders and farm rubbish, and is impenetrable. The National Trust are major landowners, and sensitive about the well-being of these caves, which are regularly monitored.

CHESHIRE WOOD CAVE GR113533 is situated below rocks at the top of a sweeping escarpment, in a wood facing north towards Beeston Tor, with panoramic views of the valley below. It is difficult to locate, and well protected by sheep-proof wire fences on land not open to public access, although visited by local caving clubs. It has a low, drystone entrance wall, and although only a forty-feet-deep crawl, has a few interesting flowstone formations. Early Neolithic remains of four people, including two adults and two children, were found here, along with a piece of antler, a chert flake, ancient pottery, and animal bones.

FALCON LOW CAVE GR104532 has the appearance of a burial mound from a distance, but is a rock shelter on Old Park Hill, opposite Beeston Tor high above the valley, and, like other high caves locally, was selected for burials in Neolithic times. Remains of six people were found here,

Cheshire Wood Cave is situated in a steep wood, below a rocky crest, high above the Manifold Valley.

White flowstone covers rocks just inside the passage to the inner reaches of Cheshire Wood Cave.

including two adults and four children, deposited with flint flakes, Neolithic and Bronze Age pottery, a deer antler, and animal bones.

ST BERTRAM'S CAVE GR106540 lies at the foot of the cliffs of Beeston Tor, partially obscured by trailing ivy, in a shadowy cleft. To reach it when the river is running in spate, before the dry season, is an adventure of wading and scrambling only spoilt by the sight of the scruffy caravans and polytunnel at Beeston Tor Farm, just across the river, when the explorer reaches this hidden corner. The most obvious cleft has a high oval window in the rock, but the easiest way in is via a narrow vertical cleft up the bank to the right, set with an iron bar to help you inside. Once inside, the chamber widens out, with a sticky floor of red cave clay, and passages branch off. A low narrow tunnel leads towards the inner chambers, which extend for 600 feet. This cave is associated with the area's own Saxon saint, Bertram, an eighth-century prince of the once pagan house of Mercia, who travelled to Christian Ireland, where he married an Irish princess. Returning to Mercia (a kingdom of the Midland counties), the young couple were forced to stop in a wood quite close to home because his princess went into labour, but while Bertram was seeking help, his wife was attacked by wolves, a common enough

Above: Beeston Tor rises above the River Manifold, and must be approached over stepping stones or along footpaths down the hill from Wetton.

Right: Crude Anglo-Saxon carvings on the font at Ilam Church record scenes from the life of St Bertram. On this panel, he brings home his young Irish wife.

hazard for lone travellers in the woods. Bertram was heartbroken on finding her bloody remains, and renouncing his princely status, vowed to become a holy man and convert the pagans to Christianity, in hopes that his loved one would be cherished in heaven, whereas she had perished unprotected on earth. Saints have to retire from the world, and the Manifold Valley became his place of refuge, the village of Ilam his place of ministry, where there is a Saxon teaching cross, and where his saint's tomb lies in the church, a goal of pilgrimage for medieval Christians. The Saxon font at Ilam church is strikingly carved with scenes from the story. He would withdraw to the cave to meditate on the beauties of creation, to mortify his flesh, and to pray. Those who wish to be alone are often pursued by agitated souls, and so it was that even after his death this place, which had obtained a holy charisma by association with the saint, was often visited by folk wanting something. Some of them may have followed the practice of leaving offerings for the saint to gain favours. A hoard of Saxon silver pennies, brooches, and gold wire were found here in one of the many subsequent digs, from the 1830s until the 1930s. The cave was used long before Bertram though, as testified to by flint tools from the Late Stone Age, a Bronze Age jet armlet, Iron Age and Romano-British artefacts, ancient pottery, animal bones, and remains of at least one, probably Neolithic, person. These can be seen in the collections at Buxton Museum.

THOR'S CAVE GR098549 is easily the most impressive cave entrance in the Manifold Valley and (with the exception of Peak Cavern) in the whole of the limestone Peak District. Its gaping mouth is visible from a considerable distance to the north of Thor's Cliff, which rises above trees just downstream from Wetton Mill. It is a steep climb up from the valley bottom, but can be tackled comfortably with a walk along field tracks from Wetton village. To get inside the cave necessitates a breathless, slippery climb up polished rocks, but once inside, amid the echoing calls of rooks and jackdaws which nest here, the large chambers, framing views of the surrounding countryside, are well worth exploring. The Midland Scientific Association, digging in 1864, found a crouched Neolithic skeleton in a stone cist, which had been buried here along with a polished stone axe, worked antler, bronze and amber artefacts from the Bronze and Iron ages, and ancient pottery. Earlier levels below the clay, which might hold clues of very ancient occupation, were not examined, as this required breaking through extensive flowstone, but other pieces of human bone were found.

Contemplative cell – the dank and muddy interior of St.Bertram's Cave.

The massive main portal (right) and West window of Thor's cave.

Daylight penetrates most of the chambers through the western portal, also known as Thyrsis's Cavern. Thor was the Norse god of thunder, who carried a gigantic hammer. How the cavern got this name is unknown, unless from the impressive echo made by striking the rock. But ingenious legends exist, including that it was formerly Thursehole, or the hole of the thurse, or fairy, or Hobhurst, after a sprite who loudly played a fiddle, which screeched and echoed down the valley. In the summer of 1870, a series of natural explosions were said to have occurred at a fissure hereabouts. Samuel Carrington of Wetton, one of the excavators at Thor's Cave, wrote to the Staffordshire Advertiser about it, claiming that three independent witnesses had heard and seen loud explosions, accompanied by flames, issuing from the rock. Previous explosions had also been heard. Carrington suggested that gas ignited by sparks from rock falls might be the explanation. He also described fountains of water issuing from the cave floor during heavy thunderstorms. In 1926, an eccentric Staffordshire man, Ralph de Tunstall Sneyd, tried to revive Druidic tradition by holding a Gorsedd of Bards at the cave. Thousands of visitors came to watch the colourful procession of druids in their flowing white robes, carrying red banners emblazoned with crosses and dragons, parading across the fields from Wetton. At the cave, they sang in Welsh and English, initiated new members, and recited poetry. After a similar ceremony at Arbor Low stone circle the following year in poor weather, and another smaller meeting at Thor's Cave, the tradition declined.

THOR'S FISSURE CAVERN GR098549, a smaller fissure cave perched below Thor's Cliff top, is reached by precariously scrambling down the rocks from Thor's Cave West Window, and traversing to the left, or from Elderbush Cave on Thor's Cliff top. These traverses are highly dangerous, and not for the unfit or faint hearted, and ropes may come in handy! Here, remains from the Later Old Stone Age were discovered. These Upper Palaeolithic hunters left worked flints associated with reindeer bones, and a large form of red deer. A dolphin bone was found, which, along with amber beads and a button, had been carried or traded from a great distance away. Later levels held the remains of six people, including four adults and two children, late Neolithic, Bronze Age and Romano-British bronze artefacts and beaker shards, and a polished stone axe. The inaccessibility of this cave says much for the fitness and stamina of these people who rarely lived beyond the age of fifty.

SEVEN WAYS CAVE GR098548 is reached by backtracking towards Wetton from Thor's Cave, then taking the stile which leads up onto Thor's Cliff

Thor's Cliff in a sprinkling of snow, with Thor's Cave (left), and Thor's Fissure Cavern above the trees in the middle of the picture, very difficult to find.

top, a steep muddy climb. Be careful when up on these cliffs not to dislodge stones, which can crash down on the unwary below. The hill abruptly turns into a precipice, and winds gust fitfully, so avoid the edges. From the highest point of the hill, the cave entrances will be seen in a rocky knoll to the south, fronting a very steep slope which leads down to Thor's Fissure Cave. The multiple entrances, including the safest from behind the knoll, are in reef limestone, and an example of a collapsed cave. Excavators in the 1950s found a few Neolithic bones had been buried, along with two leaf arrowheads. Interestingly, a pagan Saxon burial was quite close to the surface, with pottery fragments, a glass bead, and a bronze pin. The cave is in a very atmospheric position, having low natural arches, which admit gusts of wind and lofty views of the cliffs and valley below.

ELDERBUSH CAVE GR097548 lies further along the cliff top, again in a reef knoll, and, although 150 feet deep, is much shortened, and contained animal remains from interglacial periods spanning 80,000 years! Although it was said to be fully excavated, large spoil heaps remain near the entrance which have yet to be sifted using more efficient modern methods. The cave itself is curious, with a largish entrance chamber, and many small passages off. Because so much light gets inside, an alcove to the rear of the entrance chamber is filled with luxuriant ferns of botanical interest.

Seven Ways Cave, looking down at it from Thor's Cliff top.

Main passage off Seven Ways Cave, with spoil heap from diggings.

Late Palaeolithic hunters used the cave, and flint blades were found, associated with reindeer bones, including a shank worked to a point, and the remains of a cache of reindeer neck bones, showing that at a cold period of the last glacial, early modern humans were busy in the area. The remains of up to four individuals buried with Neolithic flints, including a barbed arrowhead, were also uncovered. There is evidence of use of the cave until Romano-British times, with pottery, and pebbles used as pot boilers.

DONKEY HOLE GR098550 is also known as Radcliffe's Stables after a fugitive from Bonnie Prince Charlie's Jacobite Rebellion of 1745, or the Civil War of the 1640s, depending on your choice of legend; it is about thirty feet above the river, directly below Thor's Cave, and is the smaller of the two entrances. There is some eighteenth-century graffiti, but no known ancient archaeology, possibly because the caves are so close to the river, which would have been much higher in earlier times. There is a 50 foot crawl, leading to a chamber with water, where a fugitive might hide from non-caving pursuers.

OLD HANNAH'S HOLE GR099556 is situated in a pathless wood in Redhurst Gorge, east of where the Wetton road enters the Manifold Valley. A very narrow stile hidden by wayside bushes allows access to the

GAZETTEER OF CAVES

Elder Bush Cave, hidden behind a knoll, has a large entrance.

Luxuriant ferns grow in a nook, which gets plenty of light at Elder Bush Cave.

wood, but it is not intended for public access, and there is a steep scramble up to the cave – Old Hannah must have had stout legs! Sir Thomas Wardle, a silk manufacturer of nearby Swainsley Hall, excavated the cave in 1896, on behalf of the North Staffordshire Field Club, and found human remains, flints of possibly Palaeolithic type, and Roman pottery shards. Some have disappeared, but others are at Stoke City Museum. The human bones included four adults and one child of Neolithic or Bronze Age periods. Trevor Ford, in his book on Peak caves, describes the cave as a ten-foot-high fissure, and narrowing over its forty-five foot length, situated on the west side of Redhurst Gorge. He states that this cave and Darfar Ridge Cave nearby were the sites of supposed natural explosions.

Old Hannah's Hole – contemplating the remains of a dead sheep!

Above: Difficult to get to, and hardly worth seeing – Darfar Ridge Cave securely sealed.

DARFAR RIDGE CAVE GR098558 lies just below the crest of a small but steep knoll, facing away from the river and towards Wetton Hill, above a dry valley entered over a stile from a large lay-by off the Wetton Road. The cave has a small entrance among bushes, and is sealed with a steel door. Excavations revealed a leaf-shaped arrowhead, microlith, thumb scraper, and Romano-British to medieval artefacts, along with the remains of one possibly Neolithic burial. Don Bramwell said lower levels held reindeer, and other late glacial vertebrates.

OSSOM'S CRAG CAVE GR095557 held a piece of human bone radiocarbon dated to 4,860 years old, placing it in the Neolithic period. But a dig in the 1950s revealed much activity by late ice age reindeer hunters in the Palaeolithic who brought reindeer, wild horse, red deer, and bison carcases to the cave thousands of years before that. The cave is on the opposite side of the Manifold from Darfar Ridge, at the foot of Ossom's Crag, and to reach it you must climb down to the river below Darfur Bridge, and up the steep wooded slope to the foot of the crag, where in the bushes an interesting cave, with an upwards fissure and a downwards snaking passage, will reveal itself. Don Bramwell said that the Late Palaeolithic hunters manufactured flint implements on a ridge-shaped anvil stone set in the floor, but if so there mustn't have been much elbow room, as the

Vertical fissure leads up inside the cliff from the entrance to Ossom's Cave.

Ossom's Cave zigzags downwards. Reindeer hunters knapped flints and camped here.

passage is very narrow, and winds about a lot. Outside is an area where fires would have been lit and most flint knapping and other work would have taken place. Spoil heaps from the dig litter the slope. In the crag above is Ossom's Eyrie, a small archaeological bone cave and a former lynx den accessible only to climbers.

WETTON MILL ROCK SHELTER **GR095561** is on the north side of Nan Tor Rock, the perforated limestone eminence above the bridge by Wetton Mill tearooms. The land is owned by the National Trust but tenanted, so observe courtesy if visiting. Nan Tor Cave faces the valley, and is potentially archaeological but has never been excavated. It is riddled with holes 115 feet deep, and one can emerge in unexpected places! The rock shelter, which *has* been excavated, revealed use from late glacial times, and the bones of glutton, arctic fox, and lemmings. It also contained the bones of four people including one adult and three children of Neolithic or Bronze Age date. There were also Mesolithic, Neolithic, and later flint and bone tools and pottery fragments from the Stone Age right through to the seventeenth century. Animal bones from Mesolithic levels were radiocarbon dated to 8,847 years ago.

A tea break on the Manifold Trail at Wetton Mill, with Nan Tor rock and cave above. The chocolate cake is delicious.

MILL POT **GR097561**, up the hillside, above the buildings at Wetton Mill, near the woods, a vertical pothole, is a reminder of when the river flowed at a much higher level than today. It contained an assortment of human and animal bones, an urn, and an unusual beaker, and may have been used for burials by dwellers in the rock shelter nearby.

ECTON HILL CAVES **GR102570**; of these, Sycamore Cave, on the east-facing slope of a hill where copper was mined in the Bronze Age, yielded a copper axe, pottery, flint scrapers, animal bones, and the remains of two human infants. There is a 3,500-year-old antler pick at the mining museum in Matlock, recovered from the copper mines here at Ecton, and spoil heaps from the operations, which were intensive in the eighteenth century, litter the hillside, which is riddled with up to forty dangerous, and very deep, shafts. The cave entrance is small, and marked by a moderate-sized tree. There are no paths to the cave. The Ecton Mine Educational Trust organise visits to the mine, and their field study centre on the other side of the hill.

Looking down the Manifold Valley from Thor's Cliff top, with Ecton Hill to the extreme left.

CENTRAL AREA

Lathkilldale is a long, sinuous valley, winding through the limestone plateau from west to east, where it eventually joins the Wye below Haddon Hall, shortly before it flows into the River Derwent. Walking from the medieval lead mining village of Monyash, one enters a dry valley, because the waters are rushing underground from Knotlow Cavern, to emerge, in wet weather, from Lathkillhead Cave as a river surging over large boulders. The limestone walls of the valley first narrow into a rocky ravine then widen to reveal a grand prospect of high white cliffs, whose grassy slopes are decked with rare flowers in the spring and early summer. Somewhere near here, a cave yielded the remains of three people, but its exact whereabouts are unknown. Lathkillhead Cave is passed on the left, and soon a footbridge leading across the infant river into wooded Cales Dale to the south is reached.

ONE ASH CAVE **GR172651** is situated to the right of the point where Cales Dale narrows to a rock walled ravine. You will reach it by crossing the Limestone Way footpath, ignoring the upward path to One Ash Grange, and heading on up the vale until a wall with a wicket gate blocks the valley,

Lathkilldale in winter at the point where upper Lathkilldale starts to widen out.

Lathkillhead Cave; sometimes a rushing river in wet weather – here completely dry.

and a small cave is visible in the rock to the right. One Ash Cave is just beyond this, up an earthen bank, on a shelf hidden by bushes in an angle of the cliff. Neolithic bones were found here, along with a discoidal knife, leaf-shaped arrowhead, and flint flakes suggesting usage by Late Palaeolithic reindeer hunters. Most of the caves in this valley are active badger setts, and are surrounded by loose earth, which is dirty and slippery.

CALLING LOW DALE GR183654, a virtually inaccessible side valley at most times of the year, is more so in the summer when the treacherous moss-grown limestone rubble and fallen trees underfoot are shrouded in an overgrowth of dense nettles, some reaching six feet or more in height due to the poor light penetrating the shady tree canopy. So, if you get permission from the farmer to climb his stonewalls, and don't mind being repeatedly stung or twisting your ankles, this is the place to be! These nettles easily penetrate thin fabric. Best approached over the fields from the Limestone Way footpath, where it crosses near Calling Low Farm, you will enter Bee Low Wood at the head of the valley. Slow and careful progress through undergrowth more reminiscent of a jungle adventure will bring you to an amphitheatre of rock, and up against the crag on the right-hand, eastern side is a rock shelter with a low drystone wall in front. Here, no fewer than fifteen to sixteen individuals were buried, both adults

Inside the entrance of the small cave in Cales Dale pictured in the last photograph of the colour section.

The rock shelter at Calling Low Dale, guarded by nettles and a drystone wall, hid many skeletons.

and children, beneath the dry overhanging rock face. Flint and chert patinated blades associated with horse and red deer bones suggested Mesolithic activity, and there was a petit tranchet derivative arrowhead of a similar age. Later pottery suggested Neolithic and subsequent use. The last visits, judging by decayed ground sheets and rusted lager cans, seem to suggest a 1970s-80s date!

Many more natural caves in Lathkilldale hold evidence of early man, but have yet to be explored. Lead mining has taken place here, and the remains of a mine manager's house, built across the entrance to one of these shafts, are still visited by cavers. In the 1850s there was a brief gold rush, when small amounts of the yellow ore were discovered in mineral veins, but proved not to be in commercially viable quantities.

DEMONS DALE CAVE GR168704 is difficult to identify at this grid reference. There is an obvious resurgence cave just off the footpath from the car park above Taddington Dale, visible as one heads south-west for Dimmins or Demons Dale, but it seems unlikely to have been used for habitation, as it is clearly a seasonal watercourse and full of jagged boulders. However, Demons Dale Cave is also listed as Taddington Dale Resurgence Cave, with remains including a flint dagger, chert artefacts, BA and RB pottery, and bones of four Neolithic and Bronze Age people

Demons Dale narrows to a stream and mossy rocks.

Small double entrance to rock shelter up a slope in Demons Dale.

held at Sheffield City Museum, so this may well be it. Further on, cross a stream, and after checking the information board you can enter Demons Dale, and in the cliff to the right is a small rock shelter up a slope. This may be the rock shelter excavated by Armstrong in 1948, revealing Neolithic to Roman remains. Horsborough Romano-British settlement was on the rocky rising ground just to the south of Demons Dale, and even today the area has a strange, uncanny atmosphere, with its rock gulley entrances, humps and bumps, and oddly-shaped trees and bushes. Because of the lie of the land, it is easy to feel disorientated here.

OLD WOMAN'S CAVE GR165708 is above the crags in Taddington Wood, about 100 feet above the road through Taddington Dale. There is no proper path from the car park, and to find the cave is a scramble through broken ground and undergrowth. Once found, it is an easy climb fifteen feet down into a chamber, from which small passages lead off for about 50 feet. Two Iron Age knife blades were recovered from this cave, along with pottery of the period.

HOB'S HURST HOUSE GR175712, on the north side of the steep Fin Cop hill-fort / settlement, is a narrow descending fissure behind landslips in the rock. It is best reached from Monsal Head car park, along a footpath which descends then a sub-path branches off to follow the valley side.

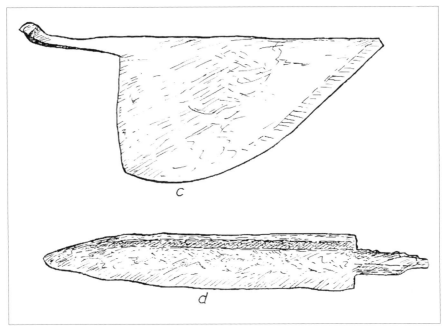

Broad and narrow knife-blades of Iron Age date, from Old Woman's Cave.

The skeleton of a teenager was found, crouched in debris with remains of Bronze Age pottery. A local legend speaks of 'a gigantic being' which haunts the valley associated with this cave.

RAVENCLIFFE CAVE GR173735 was visited from Mousterian times onwards, judging by remains, and held the skeletons of over twenty people of late Neolithic date, two polished stone axes, flint flakes, a leaf-shaped arrowhead, and pottery down to Romano-British times. The Mousterian flint scraper of 30,000 – 40,000 years ago, now in Buxton Museum, was found here, showing a very early use of this cave comparable only with Creswell. A pair of pleated thin sheets of gold jewellery from the Bronze Age was also recovered. The cave is very inaccessible, and is situated in a special nature reserve managed by English Nature, and permission is required to visit. The crags are occasionally used by climbers, and are quite dangerous, as the steep scree slopes of broken limestone rubble, saplings, and thick thorny undergrowth make this a no-go area for the unfit. Beyond a narrow fissure at the cave entrance, there is a single large chamber, with a very broken floor due to excavations, and a crawl at the back. In front of the cave is a south-facing area, with a retaining drystone wall. The cave is tucked into a fold of the crags, and quite invisible from above and below. However, the crags may be comfortably admired from the footpath, which passes Ravensdale Cottages in the valley below, approached from the village of Cressbrook.

Rubble-strewn interior of Ravencliffe Cave looks out onto a sunny Stone Age patio.

Successive digs have removed most of the cave earth, leaving a very uneven surface.

Ponies graze precariously on the cliffs near Ravencliffe Cave.

The views from above Ravensdale towards Peter's Stone and the beautiful wide-open expanse of limestone country, bordered by drystone walls, between Wardlow Hay and Eyam Edge, is one of the best Derbyshire has to offer, and one of the finest in England. An artist wrote: 'On a summers day with heat haze quivering in the air, the land to the north of the Hay Cop, or south of Eyam Edge, seems to lie in a shallow bowl giving the feeling of something very beautiful and fragile and precious held in the hollow of one's hand.' This idyllic scenery is the start of the descent into Middleton Dale, leading down to Stoney Middleton, below the plague village of Eyam, where the villagers nobly sacrificed themselves in the mid-seventeenth century, under the leadership of the vicar, by cutting themselves off so the disease would not spread. Middleton Dale is thickly wooded nowadays, with the towering limestone cliffs above rising, chimney-like, here and there through the trees. There are still remnants of industry, with a working quarry and industrial units in old mills, but they are well hidden by the natural beauties of the area, which is a paradise for climbers and cavers, while walkers and tourists visit pretty Eyam village up the hill.

MIDDLETON DALE CAVE doesn't have an exact grid reference listed, but may be near the many fissures and entrances in the vicinity of Carlswark Cavern at GR221758, on the north side of the dale (there are similar small caves higher up the dale, beyond the turn-off for Eyam). This cave is reputed to have revealed the bones of twelve or more individuals excavated in 1934, but what happened to them, or whether they were Neolithic is unknown- if they were earlier it would be very unusual. My favourite contender for this cave is situated on a flat ledge of rock with a sheer drop, halfway up a cliff facing the Rock Mill Business Park. It goes in about 20 feet, then turns sharply to the west, and contains red cave earth, which seems to have been dug away to the bare rock in the entrance passage. This is not a route for those who suffer from vertigo.

Carlswark Cavern is an extensive system, with a cockle passage showing large fossil shells in the roof. It once had a long arched entrance, but was filled with rubbish until only a creep-hole remained. In the eighteenth century, a pedlar attending Eyam Wakes was murdered at the Man in the Moon pub in Stoney Middleton, and his body was hidden here for twenty years. Rediscovered, his bones were taken to Eyam church where they remained on display, and the bell-ringers used to wear his buckled shoes. How we advance in sensitivity! Carlswark Cavern was connected to Merlin's Cave GR218759, a former show-cave which became flooded, situated in the pretty wooded dell known as The Delph, which leads down from just below Eyam Hall in the village. In the summer each year, a service of commemoration for the plague victims is held beneath a natural arch in the limestone at the north end of The Delph, known as Cucklet Church Cave.

The cave is entered from a ledge high on a cliff face in Middleton Dale.

A bit Tolkienesque – an entrance to Carlswark Cavern via the roots of a tree!

The vicar held open-air services here during the plague, in hopes of limiting the spread of infection amongst his congregation. Of the many caves around Middleton Dale and Eyam, few are in their original form because of the stone quarrying and mining, which has altered the rock faces.

HARTLE DALE CAVES **GR165803** are situated in a small dale on the west side of Bradwell Dale, and can be approached by a footpath from Hartlemoor Farm, or up the dale from Hazelbadge. At least three caves, about fifty to sixy feet deep, and a fissure yielded archaeological deposits. They were excavated by Rooke Pennington in 1875, who found Pleistocene mammal remains. The valley is very overgrown, and the caves are situated on ledges of rock in the cliffs, with access impeded by bushes. Visits to these caves, even by cavers, have been discouraged by the landowner in recent years.

Bradwell Dale has many caves, including the former show-cave, Bagshawe Cavern GR172809, found in 1806 by lead miners. This extends for a third of a mile, and has many beautiful formations and a shining, white stalagmite flow chamber, but the finest formation is hanging from a perfect, natural, domed roof seen only by cavers – a single, very long, pure white straw stalactite, whose beauty is said to make you catch your breath.

A ledge leads beyond the ivy bush to a cave entrance in Middleton Dale.

On the other side of the dale is Hazelbadge Cave, much modified by miners, and not far from a Bronze Age burial mound on top of the hill.

ROBIN HOOD'S CAVE **GR243837** on Stanage Edge, still used by climbers for camping, is an unusual wind-eroded hole in the gritstone escarpment, associated in legend with the famous medieval outlaw, whose lieutenant, Little John, is said to have been a native of the nearby village of Hathersage, and is reputedly buried in the churchyard, in a grave still tended by the Ancient Order of Foresters. This grave was excavated in the early nineteenth century, when a thighbone of massive size was uncovered. A steel cap and longbow were once displayed inside the church as having belonged to the village's mighty son. The cave and the church below are well worth visiting. Climb to the highest point of Stanage Edge, and look for a scramble down to a lower ledge, where among jumbled boulders you will see a back way into the cave. Avoid the front entrance, as this necessitates a very hazardous skip over a chimney in the rock, falling sheer for fifty feet. The cave consists of two chambers; the first, with a commodious external balcony beneath an overhang of rock, seems almost perfectly spherical, with much evidence of soot blackening of the roof by campfires, and the graffiti of generations.

Above: Robin Hood's Cave, Stanage Edge, with its overhang and rocky ledges.

Left: Inside Robin Hood's Cave, where many fires have been lit, and many initials inscribed.

The rear chamber, which provides the safest access into the cave, is crouched and dirty for reasons best left to the imagination! Other pockets of soft sand in the harder grit are still being eroded in the adjacent rocks, some forming bowls which fill with rainwater, others the beginnings of small caves. It is easy to see why this cave would provide a good hideout for outlaws, being almost invisible from above and below even to those who know of its existence, yet providing an excellent lookout point. Doubtless it would have been utilised by early man.

Cave Dale Cave GR149826, below the rear of Peveril Castle's Keep, was excavated by Pennington in the 1870s, when he removed the bones of one Neolithic person, and recovered worked bone and antler, pottery, flint artefacts, and a bronze axe. Enter Cave Dale from near the Market Square in Castleton village, via the impressive natural limestone 'gate', where an information board tells you that this was once a reef lying beneath a tropical sea, and that sharks would have been swimming high above your head in sunlit seas, with waving fronds attached to the limestone walls. Then the dale widens out into an interesting dry gorge, with a stream higher up, which suddenly disappears into the depths of Peak Cavern beneath your feet, where it forms the backdrop to Roger Rains House. Cave Dale Cave is one of several in the valley, but is situated up a steep slope, at the foot of a crag under the castle keep, which looms above. A bedding cave eight feet wide and only a few inches high, filled with soil debris, it goes in for about

The first archaeological cave in Cave Dale is on the right as you enter the dale.

Above: A bone-cave packed with badger earth high on the slopes above Cave Dale, but below the castle rock.

Right: Peveril Castle Cave is behind the iron railings immediately beneath the castle keep.

fifteen feet, and affords a fine view of the dale from the entrance. Two other caves nearby also had archaeological remains, including the twenty-feet-deep Peveril Castle Cave in the foundations of the keep, and which once reputedly connected with other caves nearby. This is totally inaccessible, dangerously situated, and can only be glimpsed from adjacent crags.

TREAK CLIFF SEPULCHRAL CAVE GR135829 was on the hillside above Treak Cliff Cavern, near Castleton, but has been obliterated by fluorspar mining operations. Excavated by Favell and Armstrong in 1921, it revealed a polished stone axe, antler pick, flint pebble, and the bones of three or more Neolithic people. Some of the bones can be seen in a case at the Treak Cliff Cavern visitor centre and café. A path behind the centre leads on up the steep hillside below the fluorspar workings, and eventually onto the ridge of reef coral above.

WINDY KNOLL GR126830 is situated on a reef knoll to the South of Mam Tor, not far from a footpath, though you have to search among the hummocks to find it. The original fissure, which yielded thousands of Pleistocene animal bones, including a 37,300 years old radiocarbon dated bison bone, has been removed by quarrying, but a large cave, 120 feet deep, can be seen, which has recently suffered a collapse near the entrance and is closed to visits, and now provides undisturbed nesting to swallows. Pennington, Dawkins, and Tym excavated here from 1870-76, and discovered that a waterhole with a hidden fissure had trapped many ancient animals. Some human bones are listed as having been recovered, but whether from the fissure or the cave is unclear. All finds have been spread around a variety of museums, including Buxton, Derby, Cambridge, Sheffield, and the British Museum. Curiously, Stockport Museum holds Neolithic human bones said to be from Windy Knoll, acquired in the 1940s, but who donated them is a mystery.

THIRST HOUSE CAVE GR097712; best approached by a track, then footpath from Chelmorton, rather than past the dusty quarry workings at the foot of Deepdale. This is one of the two 'Deepdales' running south from the Wye valley, and lies to the west of Chelmorton, while the other is to the east of Taddington. The village of Chelmorton has the finest remnant of medieval field strip farming to be seen in the Peak, a squat medieval church, the highest in Derbyshire, with remains of a 'Llan' original enclosure, and carved tombstones of ancient foresters in the porch. The village was formerly ditched like Castleton, but whether to keep animals in, or wolves and outlaws out, is difficult to say. There is a village spring, and two burial mounds on the hill, the Lord and Lady's. Thirst House Cave lies in a valley reputed locally to be the haunt of fairies. Excavated by Salt and Ward in 1884-1900, it yielded many Romano-British artefacts now in Buxton Museum, and the remains of at least four people. One skeleton was found

Impressive entrance to Thirst House Cave, Deepdale, said to be a burial place of Brigantian chiefs.

Looking back towards the entrance across the uneven floor from the inner chamber in Thirst House Cave.

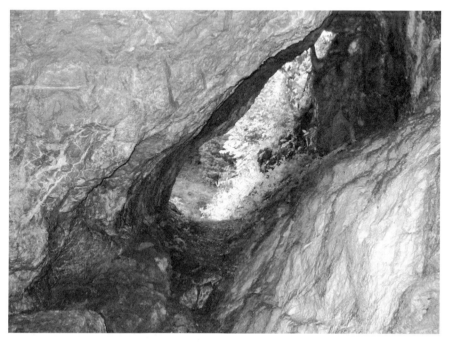

Marbling effect of the rock in small cave near Thirst House, beside the footpath into Deepdale from Chelmorton.

in a shallow stone-lined grave in front of the cave, with an iron spearhead. The cave has an impressive entrance, and is on the east side of the valley above a stone lip. There are two chambers, and the cave goes in for 190 feet. Watch out for the uneven floor, and a hole which descends among boulders to a rift with a pool. There are other small caves close by, and across the dale is Deepdale Cave, a seventy foot bedding crawl.

EASTERN FRINGES

Ash Tree Cave **GR514761**, near Whitwell, on the borders of Derbyshire and Nottinghamshire, lies in magnesian limestone in rolling countryside, and was excavated by Armstrong 1938-57, then West and Riley 1959-60, and revealed remains of seven people, including five adults, one child, and one cremation. Most activity seems to have been Bronze Age and Romano-British, and charcoal, flint flakes, an arrowhead, and pottery were found. A human bone was dated to 3,730 years old.

Various caves in the magnesian limestone show human activity, most notably in and around Creswell Crags, and at Robin Hood Cave and Mother Grundy's Parlour, which both revealed human remains of Neolithic date, and much, much, older artefacts (see earlier chapter). In addition, Langwith Cave at Upper Langwith GR517694, had remains of two people of Iron Age date, flint artefacts, and animal bones; Steetley Quarry Cave, near Shireoaks GR 558811, had human and animal bones; Thorpe Common Shelter, near Thorpe Salvin GR528793, unusually had one person of Mesolithic date with chert and flint artefacts, and radiocarbon-dated charcoal 6,600 years old; while the rock shelters excavated at Whaley, GR515717 and 511721, contained two adults, a greenstone axe, and flint and quartz artefacts suggesting use back to Mousterian times.

Chelmorton village, with its medieval strip fields, drystone walls, and ancient church, nestles below Chelmorton Low topped by burial mounds.

CONCLUSION

Derbyshire was just southerly enough to escape the absolute devastation of the last ice age to afflict Britain, about 20,000 years ago, while that of 450,000 years ago had entirely covered the area with ice a mile thick, and much of the rest of the country also. This explains the survival of ice age art at Creswell, and evidence of Old Stone Age culture in the Peak. Humans had been around throughout all these epochs, either here or in Europe, and the latest research by the Natural History Museum's AHOB team suggests Britain was repeatedly colonised then abandoned by pre-Neanderthals, Neanderthals, and finally early modern humans at least seven times, as the ice advanced then retreated in the last 700,000 years. Caves were naturally best suited to conserve archaeological deposits throughout these periods, despite the addition of glacial sludge and wind blown sand. During that time, very minor changes were made in the tools and ways of hunting and gathering adopted by human groups, by comparison with the staggering rate of change accrued among modern humanity in the last few centuries. An expert was asked if he thought early man had language, and his reply was that *if so* he must have been saying the same thing over and over again for hundreds of thousands of years!

What has happened to the human brain to make us so adaptable that we have increased from a few million specimens worldwide a few thousand years ago, to the billions of today? Analysis based on early history might suggest that our 'intelligence' depended mainly on a learning resource, passed on from generation to generation, with a limited potential for innovation. Those previous peoples, including the prototype *Homo Erectus*, lasted successfully for up to a million years in the face of climatic and other disasters, but ultimately all failed. The occupation of Britain by returning ice-age hunters, beginning about 11,500 years ago, continues to the present day, in one of the longest warm phases between glaciations we have yet experienced. Modern Humans, such as the Gravettians of 27,000 years ago, had fared better than Neanderthals who previously survived for hundreds of thousands of years, including the harsh climatic epochs. The sudden disappearance of Neanderthal from the archaeological record coincides with the appearance of this new 'top predator.' Cut marks on bone from Gough's Cave suggest cannibalism among modern human populations. Was this the ultimate fate of Neanderthal too? Future research may throw light upon this, and the still-powerful human concept of 'sacrifice.' Of the hundreds of caves in Derbyshire, only a fraction have been properly investigated, and a re-analysis of the finds from Creswell by Roger Jacobi of the Natural History Museum and others are leading to exciting new areas of exploration. A cave as rich with potential as the massive Peak Cavern has never experienced a major excavation, and it would be strange indeed if the glacial terraces there did not hold fascinating discoveries for the future.

MAPS

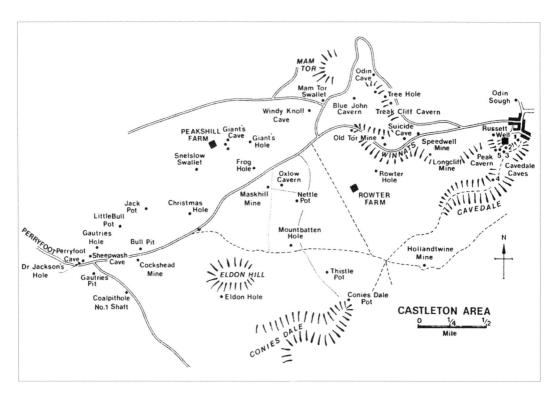

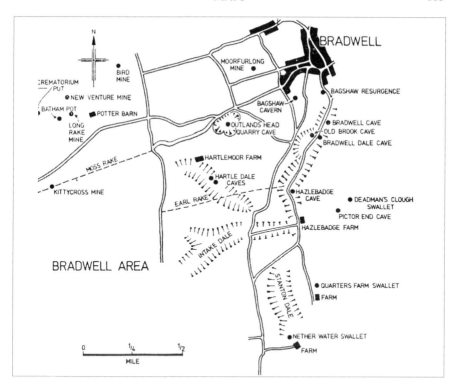

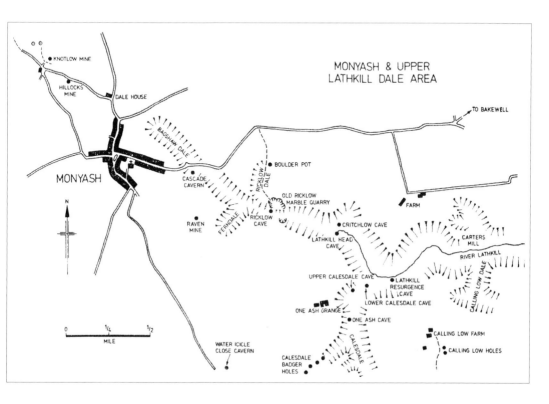

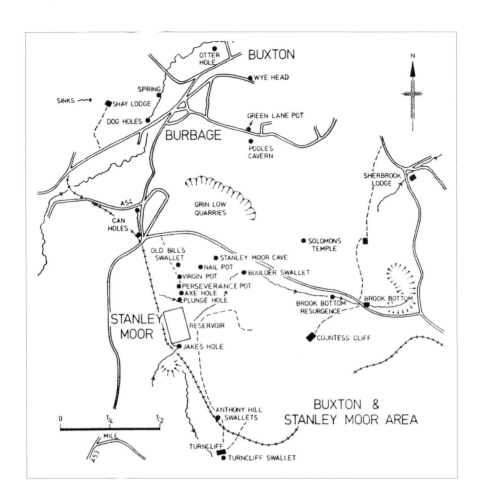
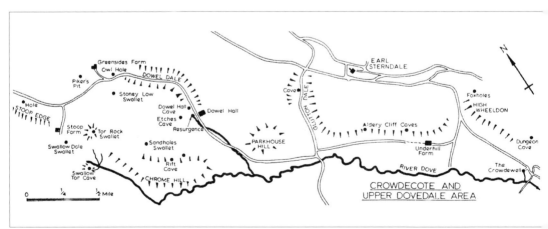

MAPS

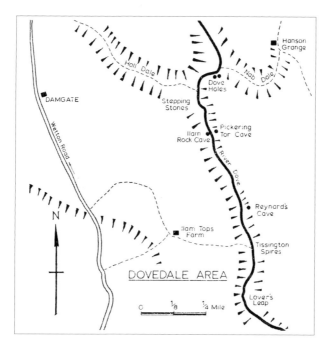

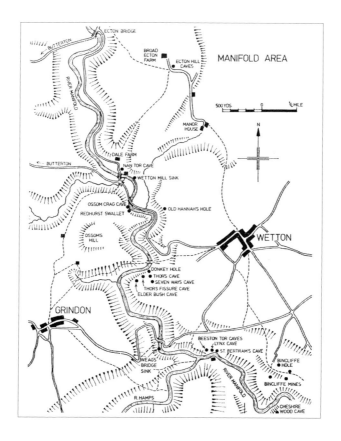

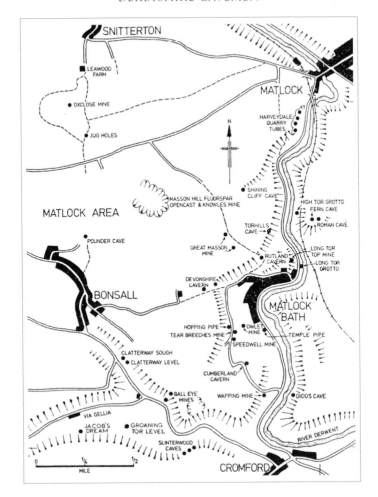
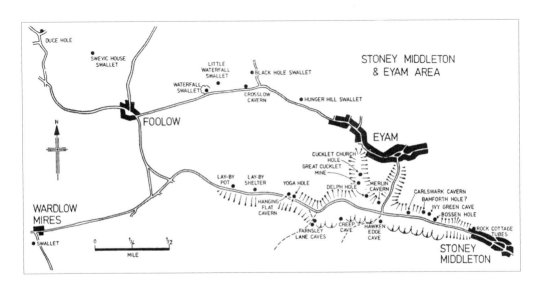

BIBLIOGRAPHY

A Conservation Audit of Archaeological Cave Resources in the Peak District and Yorkshire Dales, University of Sheffield Research School of Archaeology, 2006

Bahn, Paul & Pettitt, Paul, *The Ice Age Cave Art of Creswell*, English Heritage, 2009

Baker, Ernest, *Moors Crags and Caves*, John Haywood, 1903, Halsgrove reprint, 2002

Barton, Nick, *Ice Age Britain*, Batsford & English Heritage, 1997, revised 2005

Bateman, Thomas, *Vestiges of the Antiquities of Derbyshire*, John Russell Smith, 1848, reprint Scarthin Books

Bramwell, Don, *Archaeology in the Peak District*, Moorland Publishing Co., 1973

Bramwell, Don, *Limestones and Caves of the Peak District*, Geobooks, 1977

Boyd Dawkins, William, *Cave Hunting*, Macmillan 1874

Brighton, Trevor, *The Discovery of the Peak District*, Phillimore, 2004

Chamberlain, Andrew, & Williams, Jim, *Caves Fissures and Rock Shelters in the North Midlands*, Dept of Archaeology & Prehistory, University of Sheffield, 2001

Creswell Crags – a guide to the caves and ice age remains, Creswell Heritage Trust, 2009

Derbyshire Archaeological Journal, archive at Buxton Museum

Ford, Trevor, *Caves of Derbyshire*, Dalesman, 1974

Ford, Trevor, *Castleton Caves*, Landmark, 2008

Ford, Trevor, *Treak Cliff Cavern and the Story of Blue John Stone*, Treak Cliff Cavern, 2004

Ford, Trevor, *The Story of the Speedwell Cavern*, Speedwell Cavern, 1990

Kirkham, Nellie, *Derbyshire*, Paul Elek, 1947

Lawson, Andrew, *Cave Art*, Shire Publications, 1991

Lewis-Williams, David, *The Mind in the Cave – consciousness and the origins of art*, Thames and Hudson, 2004

Oppenheimer, Stephen, *The Origins of the British*, Constable, 2007

Pennington, Rooke, *Notes on the Barrows and Bone Caves of Derbyshire*, Macmillan, 1877

Peakland Archaeological Society newsletters, archive at Buxton Museum

Salt, W. H., *Discovery of Ancient Remains in Deepdale near Buxton*, Reliquary and Illustrated Archaeologist, 1897

Stringer, Chris, *Homo Britannicus*, Allen Lane, 2006

Sullivan, Paul, *Poole's Cavern & Buxton Country Park*, Jarrold Publishing, 2005

Trent and Peak Archaeological Trust, Manifold Valley Cave Survey, 1993

Various unpublished sources including notes from Derbyshire Caving Club.

ACKNOWLEDGEMENTS

I should like to thank the willing team of adventurous cave hunting photographers who risked broken ankles to obtain suitable illustrations for the book, particularly James Birch, Anthony Boardman, and Coral Dranfield, whom I must especially thank for her help in sorting out the picture files. I also wish to thank Trevor Ford for his advice about caves, and for his many authoritative books on the subject. Kevin Wilcock accompanied me on several expeditions, while old cavers Geoff Standring and Alan Burgess supplied advice, and were there in spirit. Ros Westwood of Buxton Museum helped me access their archive and fine collection of cave artefacts, and Martha Lawrence supplied information. Alan Walker of Poole's Cavern, Buxton, gave permission to reproduce excellent photos from their collection, showed me around, and was most helpful with information, as was Richard Taylor at Peak Cavern, Castleton. Creswell Crags Museum, near Worksop, was extremely helpful. The volunteer staff at Matlock Mining Museum were also very informative. A caves database by Andrew Chamberlain, Jim Williams, and colleagues at Sheffield University's Department of Archaeology and Prehistory proved a very useful guide. The late Don Bramwell wrote up lots of archaeology for the Peakland Archaeological Society, which is invaluable, as was his primer, *Archaeology in the Peak District*. Finally, if I have forgotten anyone, please accept my apologies, and the assurance that any omissions simply prove that even modern humans are fallible.

INDEX

Acheulian, 13
Alderley Edge mines, 133-136
Ash Tree cave, 113-116, 178
Arbor Low, 40, 62, 152
Armstrong, Leslie, 29, 67, 111, 127
Artists, cave art, 25, 26, 121
Axis wobble, 12

Baker, Ernest, 47, 48, 54
Bahn, Paul, 121, 124
Bateman, Thomas, 32, 33, 60-63
Basque genetics, 23
Benty Grange boar's crest helmet, 41, 62
Blue John, 68, 103-110
Boxgrove man, 12, 13
Boyd Dawkins, William, 26, 63-66, 76
Bramwell, Don, 67, 76
Bridestones, 35
Brigantia, 83, 97
Bronze Age, 38-43
Burin engraver, 25
Buxton, town, 69-71
Buxton Museum, 8, 12, 65

Calling Low Dale, 162
Castleton, 66, 88, 96
Carsington Pasture, 8, 13, 143
Cannibalism, 29-30
Cave as archetypal symbol, 25
Carlswark Cavern, 169
Cave Dale caves, 173-175
Chee Dale, 80-83
Cheddar Man, 23-24
Cheshire Wood cave, 146, 147
Church Hole cave, 121-122, 125

Clans, 29, 30
Creswell Crags, 8, 14, 17, 18, 21, 22, 27, 111-128
Cro Magnon man, 19, 20

Dark Peak, 7, 10
Darfar Ridge cave, 157
Derbyshire Caving Club, 96, 129, 133
Demon's Dale, 54-55, 164
Devil's Arse, 88
Donkey Hole, 154
Dove Valley, 136-142
Dowel Hall cave, 27, 37, 136
DNA, 23, 24
Druid's chair, 54

Early Humans, 12-18
Ecton Hill, 39, 160
Elderbush cave, 27, 153
Eldon Hole, 44-48
Etherow, 131
Eyam Moor, 33
Eyam plague, 169-170

Falcon Low cave, 146-148
Fox Hole cave, 32, 37, 38, 137
Ford, Trevor, 44
Frank in th' Rocks cave, 139

Geology, 7, 10
Gough's Cave, 23, 26, 27, 30
Gravettians, 21, 180-181
Grey Ditch, Bradwell, 36
Groups, 29

INDEX

Harborough cave, 37, 39, 51-53, 145
Hazelbadge cave, 56, 57
Hartledale caves, 170
Hermit's cave, 58-60
Hill forts 39-41
Hobbes, Thomas, 92
Hob Hurst House, 165
Holocene, 12
Hopton hand axe, 12, 13
Hope Valley, 36
Horse engraving, 27, 64, 120

Ilam village, 149
Ilam Rock, 6
Ice Ages, 11-12

Jacobi, Roger, 181,
Jet necklace, 32, 34

Kent's Cavern, 27
Kinderlow Cavern, 131

Lathkilldale, 161
Lawson, Andrew, 25
Leviathan Cavern, 103
Lewis Williams, David, 122-124
Ludchurch Cavern, 132

Manifold Valley, 7, 27, 145
Mam Tor, 36, 41, 88
Mary Queen of Scots, 77, 79
Matlock caves, 145
Mesolithic period, 29-34
Microlith, 35
Middleton Dale caves, 168-170
Modern Humans, 19-22, 181
Mousterian technology, 16

Neanderthals, 15-18
Neolithic period, 34-37

Odin's Cave and mine, 107
Old Hannah's Hole, 154
Old Woman's Cave, 165
One Ash Cave, 162
Oppenheimer, Stephen, 23

Ossom's Cave, 27, 157
Pakefield flints, 12
Paleolithic period, 13
Peak District Mining Museum, 39
Peak Cavern, 44, 51, 88-96
Peakrils, 49-51
Pegasus Caving Club, 8, 143
Pennington, J. Rooke, 65-66
Peveril Castle, 88, 91
Pleistocene, 11
Poole's Cavern, 41, 44, 71-87
Pontnewydd teeth, 13, 14
Prehistoric animals, 14

Ravencliffe cave, 14, 15, 39, 166-168
Rains cave, 144
Red ochre, 32, 33
Reynard's cave, 141
Ripoll, Sergio, 121, 127
Robin Hood Cave, Creswell, 120
Robin Hood's Cave, Stanage Edge, 9, 171
Roaches, 131

Salt, W.H., 49
Seven Ways Cave, 38, 152
Shaman, 25, 26
Speedwell Mine, 99-102
St Bertram's Cave 148-150
Stanage Edge, 171
Stringer, Chris, 124
Stockport, 129
Stukeley, William, 58-60

Thirst House cave, 48, 49, 175-178
Titan Cavern, 103
Thyrsis's Cavern, 151
Thor's Cave, 28, 150-152
Thor's Fissure, 27, 39, 152
Treak Cliff Cavern, 99, 104-109, 175
Tym, John, 66

Wetton Mill caves, 39, 159
White Peak, 7, 10
Windy Knoll, 65-66, 175
Winnats gorge, 99, 102

p35 Five Wells above Taddington
p37 Calling Law Dale